HEMP MASTERS:

Ancient Hippie Secrets for Knotting Hip Hemp Jewelry

2005
Suzi Szot :)

Written and Illustrated by
Max Lunger

Eagle's View Publishing
A WestWind, Inc. Company
6756 North Fork Road
Liberty, UT 84310

Library of Congress Number: 98-72138
ISBN: 0-943604-57-5

10 9 8 7

TABLE OF CONTENTS

ACKNOWLEDGMENTS

This book was conceived by myself, Robert L. Johnson, and John Vandermeulen. I am indebted to them for their help with the concept. Out of all the reasons for which hemp has been grown and used, I must say one of the best is using hemp to make jewelry and decorative wall or car mirror hangings.

The smiles I have seen, and the appreciation of my talent and time, which I have received from my friends and customers, have been a gift of inspiration I will never forget. To all my present and future friends and customers, I would like to say thank you.

Special thanks to: Robert Johnson, Stanley Lawrence Jr., Sherel Warwood, Jana Jones, Bob Schneiter, Lacey Morse/Lunger, Stan Brennan, Tom Eckman, Donnell Overton, Kay Reddish, Rebecca Vandevener, Mike Dyer, Willey Galiegos, Karen Huddleston, Arlin & Rondel Jones, Mood For A Day, Gail & Vicky at Beadwiser in Roy, UT, the girls at Eagle Feather Trading Post in Ogden, UT and Father Guiedow. Their helpful suggestions, insightful enthusiasm, and continued support have given me the inspiration to follow my Dream: To live life with my talents rather than my abilities. What a concept!

NOTES

INTRODUCTION

The Hemp Masters knotting guide is designed to help crafters through the process of creating hemp jewelry and other decorative pieces. Suggestions include bracelets, anklets, necklaces, chokers, car mirror charms, key chains, wall tapestries, plant hangers, speaker hangers, and hackysacks. These are just some of the possibilities. For gifts, or for profit, making hemp jewelry is a very rewarding craft.

Other mediums that are fun to use with the hemp masters knotting guide are jute, colored yarn, colored floss, horse hair, rope, leather strips, and even old clothing cut into strips.

Terms Used

Bunk (Hemp) - bad, not worthy; a section of cord where two lengths have been connected. A section like this is bunk because it usually has a knot which must be frayed out to make the jewelry look consistent (see Ancient Hippie Secrets and Helpful Hints).

Carrier or Bead Carrier - one or more cords which knots are tied around. Beads are also threaded onto these cords.

Cord versus Twine - cord and twine are wound in opposite directions when the fibers are spun. Cord is wound counter-clockwise and twine is wound in a clockwise direction.

Hip - the portion of a square knot created when the vertical cord passes over the two horizontal cords. *Or* a body part. Also someone loved by all.

Heavy - a reference to the weight or size of cords used in hemp jewelry, *or* of one's ex-mother-in-law. Profitable. Something that blows your mind.

Knotter - one or more cords used to tie a knot. *Or* a person in denial of their own actions (I did not!). Not to be confused with Knottier, which is a naval rank.

Phish (Fish) Bone - a style of knotting heavy bank jewelry (just one of Hemp Masters' best)

Sinnet - a vertical chain or braid of repeated knots.

Style - type of jewelry. *Or* a hair cut. A hip line of clothing - to be in or out of style (who cares?).

Bank - $, money, folding stuff, lettuce, long green, script, bread.

Dig - to understand or like something (I dig it.). To understand or like someone (I dig you!). What the boss tells you if you are a dirt pilot.

Freak Out - when in trouble, when in doubt, run in circles, scream and shout; a bunk expression of human emotion.

Funky - a style of music; a hair cut; something unusual; a verbal description of anything that makes you go hmmm...(that's funky!).

Hippie - a human being.

Jive - rhythmic speech, odd directions. Not to be confused with jive turkey.

Man - a slang word for the human species. Not to be confused with *The Man*.

Om - a mantric word (Hinduism). Om is where the heart is. There's no place like Om. Om my God!

It's good to be Om. A man's Om is his castle. Welcome Om.

Score - to purchase or to find. To touch home plate; get a touchdown; make a goal; keep track of points in a card game. The art of competition (Hey Man, what's the score?).

Stash - a special supply of craft materials, a special place.

Trip - to go somewhere (take a trip); to think (trip on this); to think everyone is after you (wow, he's tripped out); not understanding how something works (that's trippy).

Truck on Down - walk, run or ride to a desired destination.

Hemp Facts Everyone Should Know

Hemp has been around for hundreds of years, from ancient Egypt, to the birth of the United States of America, to the present day, and it has been used for many purposes. The U.S. Constitution, the Bill of Rights and many laws were written on hemp paper. The first pair of 501 jeans and the first U.S. flag were made from hemp.

The harvesting of hemp was deemed illegal in 1937. In 1941 Japan stopped all shipments of hemp into the United States and the U.S. government and industry changed their views on hemp (what a concept!). During World War II the U.S. government encouraged farmers to grow hemp because the navy needed rope. Every farmer in America had to sign a document stating that they had watched the government produced film called *Hemp for Victory*. Those who checked the yes box, along with their families, were exempted from the draft. Hemp, Hemp Hurrah! They had Hemp that day!

Webster's New World Dictionary defines hemp as follows: hemp (hemp), n. 1. a tall Asiatic plant of the nettle family, having tough fiber. 2. the fiber, used to make rope, sailcloth, etc. - hempen, adj. The fiber, of course, may be used to create beautiful jewelry and that's what this book is all about; teaching the reader the basics of knotting and design.

Hemp is considered by many to be the world's most versatile and valuable resource. It provides the raw material for more products (over 50,000) than any other plant. The stems are used to produce fabric, fuel, paper and other commercial products. The hemp is dried and broken down into two parts; thread-like fibers, and bits of "hurd" or pulp. From the fiber strands, which are spun into thread, come such products as the world's strongest natural fiber rope, and durable, high quality textiles of all types and textures. These fabrics can be made into sails, clothing, and fine linens. From the "hurd", which is 77% cellulose, comes such products as tree-free, acid-free paper, non-toxic paints and sealants, industrial fabrication materials, construction materials, biodegradable plastics, and much, much more.

Hemp is also one of the best sources of plant pulp for biomass fuel to make natural gas, charcoal, methanol, gasoline, or even to produce electricity.

The hemp seed is used to produce nutritional oils, lubricants and fuels. Hemp seeds are also an excellent source of protein.

Hemp foliage has also been promoted for its medicinal value in easing pain, relieving stress and treating illnesses from glaucoma to nausea in AIDS and cancer patients. Hemp roots even play an important role by anchoring and invigorating the soil, providing erosion control and preventing mud slides.

Hemp is the only plant that grows up to 20 feet throughout the United States. One acre of hemp can replace five acres of cotton used for material; over 50% of the pesticides applied in the U.S. are used on cotton crops - less than 3% kills the insects, the rest goes into the ground water or is embedded in our clothing (what a concept!). One acre of hemp can replace four acres of trees used for paper; and hemp has a three month growing season! Hemp's abundant yield can add over a trillion dollars to the U.S. economy and assure us prosperity, plenty and economic stability. These facts are substantial and make the argument for legalization of the hemp industry in the United States ... a critical one. Just think, if we grow enough hemp and save enough trees, global warming could be a thing of the past, not to mention the national debt. Hemp can save our planet!

Just a Little Hippie History

These are the knots Granny told us about!
Their ancient hippie secrets, I'm about to let out.
She told us a story, Granny spoke with a grin,
we sat in a circle, so she would begin.

She started a commune, old hippie school style.
A happy Granny Guru, they called her Granny Smile.
Mother Earth is our savior, so she defended her trees.
But a man would not listen, though she was down on her knees.
His excuse was more money, he killed the trees for his greed,
so she showed him her hemp, an alternative seed.

He had her arrested, claimed her hemp got her high!
By the time she got out, the commune went dry.
And now at the spot where she gathered her band,
the man's condominiums have infested her land.

Now she's still Granny Smiles and her vision I see,
is thousands of hemp farms, with no THC!

So please - by promoting hemp products you could save tons of trees!

HOW TO BEGIN

The hippest, most common, question asked is how long should the hemp be cut when beginning a piece. The answer is that it depends on the style, be it necklace, choker, anklet or bracelet. Say it is a necklace. Most necklaces measure 18 to 20 inches in length. The amount of hemp cord needed is two pieces, each approximately eight feet in length. Both cords will be folded in half, but not cut.

The following chart is a guide for 45 lb. test (approx. 2 mm) cord. The chart applies to jewelry with a consistent series of knotting throughout the pattern. Other factors will vary the length needed: The more inconsistent the knotting or the more beads included in the pattern, the less cord will be needed. The length of cord needed will also vary depending on the gauge (thickness) of cord used and on how tight the knots are pulled. The thinner the cord, the more knots will be tied for a given length of jewelry and the more cord will be needed. Tighter knots require less cord per knot, but more knots are needed to achieve a given length. Finally, it is always better to cut the cord too long and have some left over, than it is to cut the cord too short. Experience is the only solution (*Ommm*).

Style	Length of piece	Cut Length (2)	Folded Length (4)
Necklace	18 to 20 inches	8 feet	4 feet
Choker	13 to 18 inches	6 to 8 feet	3 to 4 feet
Anklet	9 to 13 inches	4 to 6 feet	2 to 3 feet
Bracelet	5 to 8 inches	3 to 4 feet	1.5 to 2 feet

To begin, cut two cords to the length needed for the style being made. Fold both cords in half. If any of the loop closures will be used, **do not cut** the cords (loop closures are used on nearly all the designs in this book). This gives four equal lengths of cord coming from the two loops (folds) at the top (two from each fold). For example, to start a necklace, two cords each eight feet long, would be folded in half. This will result in four cords, each four feet in length, with two loops at the top.

Place one loop inside the other and hold the four cords so that they do not cross one another. It may be easier to pin the top or fold of these cords to a craft board. The two cords from the inside loop should lie inside the two cords from the outside loop. The "outside" cords, on the left and right, are the *Knotters*; they will be the cords doing the work. The center two cords are the *Bead Carriers* or *Fillers*, around which the knots will be tied. The knotting cords will be referred to as *Right* and *Left* and are held in the right and left hand respectively (**Figure 1**). Note: The bead carrier cords do not have to be as long as the knotter cords **if the carriers will not be used for any knotting at any point** in the pattern. However, many

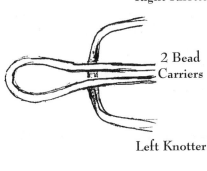

Right Knotter

2 Bead Carriers

Left Knotter

Figure 1

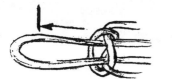

Figure 2

designs use switch knots or require the carriers to be pulled out as knotters.

Pull the inside loop in front of and above the outside loop (see **Figure 1**).

With the right and left cords from the outside loop, tie a loose Half Knot around the cords from the inside loop (see **Figure 2**). Slide the half knot close to the end of the inside loop and pull this knot tight. (See **Page 11** for directions on tying a Half Knot, if needed.)

Congratulations! The Knotters (outside loop cords) and Carriers (inside loop cords) have just been established for this piece of hemp work. Any pattern of knots described in this book can now be tied. This is also the beginning of the Slide Loop Clasp, a Hemp Masters Original (see **Page 23**) and a few other good closures for hemp jewelry.

Another way to begin knotting is to use a button (any button will do) or a flat bead. If the button has four holes, the holes should be big enough to accommodate a single cord; if the button has two holes, they must each accommodate two cords; if the button has only one hole it should accommodate four cords. If the holes are too small, *don't freak out*. There have been many devices invented that will correct this bunk hole (i.e. a file, sandpaper or saw).

The button will become the center of the pattern, with hemp knotted on either side. In this case, the two beginning cords should be cut after they are folded, resulting in four cords of equal length.

The button in the photographed example (**Page 7**) has four holes, but the beginning positions of the cords are the same, regardless of the number of holes. Thread the first two cords through the top two holes in the button positioning the button in the center of the cords. (For a two hole button, thread these two cords through the top hole and for a one hole button, position them at the top of the single hole.) Fold the cords in half and to the top, catching the button in the loop this creates.

Thread the second two cords through the bottom two holes in the button and fold them in half to the bottom. (For a two hole button, thread these two cords through the bottom hole and for a one hole button, position them at the bottom of the single hole.) This gives four strands at the top of the button (two strands from each cord; one becomes a knotter on the outside and one becomes a carrier on the inside) and four strands at the bottom of the button (again, knotters on the outside and carriers on the inside). This is shown as **Step 1** in the first photograph.

Tie a Half Knot on either side of the button to secure the button and all the cords in position (see **Step 2** photo). Tie the same pattern on either side of the button and this piece of hemp jewelry will be done. The third photograph on **Page 7** shows the cords threaded through a single hole flat bead, with work begun on the first half of the piece.

Ancient Hippie Secrets and Helpful Hints

Buying Hemp - Notice the gauge or test weight of the hemp. The author's favorite sizes are 20 lb. (about 1mm), 45 lb. (about 2 mm) and 170 lb. (about 4 mm) test. Also available is 80 lb. test (about 3 mm). Also notice the clarity or how tight and clean the cord or twine is wound. If the cord is too fuzzy, beeswax is a great conditioner.

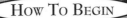

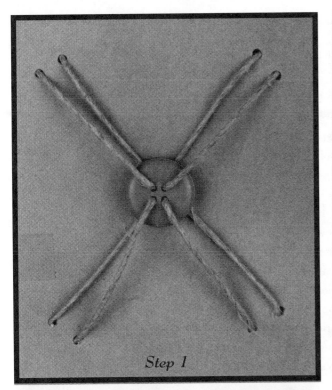

Step 1

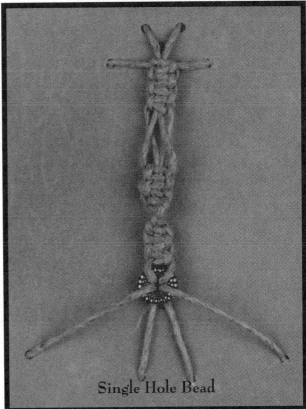

Single Hole Bead

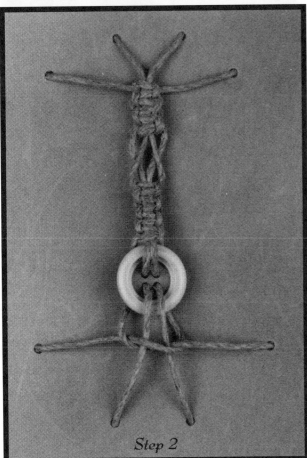

Step 2

Growing Hemp - Don't do this in the USA (it is still illegal in most states)!

Colored Hemp - Colored hemp is made a bit fuzzy and weaker by the process used to dye it. Therefore, it works best if used in conjunction with stronger, naturally colored hemp. Use beeswax to eliminate the fuzziness and take care not to pull the colored hemp too hard, or it might break. If it does break, undo the last four or five knots and add a new colored piece.

Tying All Knots - Before pulling any knot tight, twist the hemp cord in the same direction that the cord is wound. This is important for maintaining the shape of the knot and helps define the more complicated patterns. This technique is especially helpful when using Alternating Square Knots, Butterfly Knots, Pretzel Knots or loops of any kind.

Adding Cord (s) -

Knotters Are Too Short - If the knotters get too short to finish a piece, *don't freak out*. Cut a piece of cord long enough to finish knotting the pattern; when deciding the length, keep in mind that this cord will be folded and become both knotters. This technique can also be used to add extra cords and hide them inside the knotting.

Place the middle of the cord to be added on top of and across the bead carriers. Pull the existing knotters over this new cord and tie a Half Knot around the bead carriers. Be sure to pull this half knot snug around the new cord being added; this knot hides the addition of the new cord and helps to secure it (see **Figure 3**).

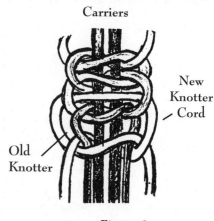

Figure 3

Pinch the short, old knotters parallel to the bead carriers. Use the ends of the cord just added as the new knotters and tie a Square Knot around all four of the old cords - allow the short old knotters to become carriers for the time being. Tie at least three more tight Square Knots with the new knotters - this holds all the cords securely in place. Glue (not super glue) also helps. Cut off the old knotters below the square knots just tied.

Carriers Are Too Short/Need Another Carrier - If the carrier(s) get too short to finish a piece, *again, don't freak out*. One or more new carrier(s) can be added at the same time to replace the short ones. This technique can also be used to add an extra carrier. Cut one or more cords to the length needed.

Place the new cord next to the short carrier(s). Pull the cord to be added at least one inch above the last knot tied. With the knotters, tie a Half Knot around all the carriers, new and old. Take the one inch piece of new cord, which is above the knot, and fold it down over the half knot, so it lies next to the other bead carriers (see **Figure 4**). Tie three or more tight Square Knots to hold everything securely in place, then cut the short carrier(s). A small amount of glue on the carriers and square knots will also help hold them in place.

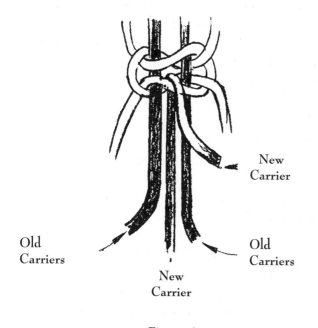

Figure 4

One Knotter is Too Short - If only one knotter is too short, this too can be corrected. Simply add one new carrier as just described. Once the new cord is secured, bring the new cord out as a knotter. At the same time, bring the old short knotter to the center as a carrier until it runs out or is cut off.

Adding Extra Cords (A Masters Trick) - Some more advanced patterns require the addition of extra cords. For instance, this technique may be used to go from a single knotted portion to two connected portions which are the same thickness as the original as shown in the Car Mirror Charm on **Page 81**.

To add extra cords, lay them across the carriers and put the knotters on top of the cords being added. In the **photograph** the cords labeled 1, 2, and 3 are new and lie in front of the carriers. New cords can be added either in front of or behind the carriers. Tie a Half Knot tight against the added cords (**Step 1**), then tie three Square Knots to hold the added cords in place. Leave the new cords out to the side or pinch them in with the carriers (**Step 2**) depending upon the pattern being tied. This technique is particularly useful for Phish Bone and Alternating Square Knot patterns.

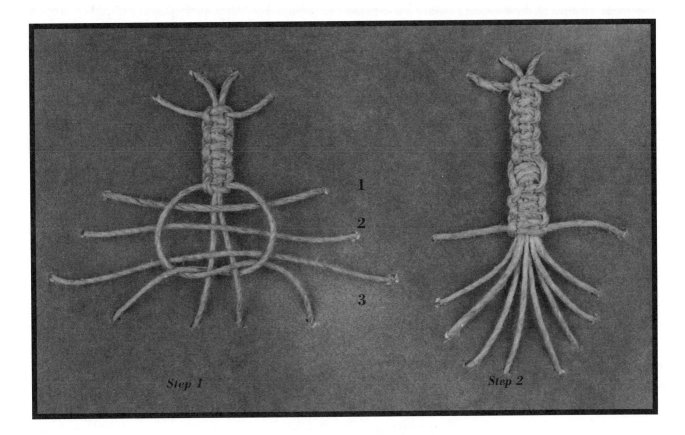

Step 1   1   2   3   *Step 2*

Preventing Knot Slippage - This technique will stop knots and beads from slipping along the bead carriers. Tie a Half Knot with just the carriers, then use the knotters to tie a series of knots over the knot in the carriers.

Threading Small Beads - To get small beads to thread more easily on hemp cord, put some glue on the cut end and twist the cord as it dries. If the bead hole is very small, cut the end of the cord at an angle and twist the end into a point as the glue dries. <u>Do</u> <u>not</u> <u>use</u> <u>super</u> <u>glue</u> as it makes the hemp brittle and causes it to break. Rubber cement is the best *score*.

Bunk Hemp - No matter how good the roll looks, there will always be a "bunk" spot where the bead stops. To correct this headache spot, lick the hemp or get it wet and scratch it to make it fray. Then cut the frayed edges until the bead passes over the spot. Too much and the cord will break! Glue helps keep the cord wound.

Dirty Jewelry - Wash your hands! Or wash the jewelry using warm water, shampoo and conditioner. Air dry. Washing weakens hemp over time, but many people wear simple hemp jewelry patterns all the time, including in the shower. Complicated patterns do not fare as well when washed (see wet hemp), and this should be avoided.

Wet Hemp - Hemp swells when it is wet; the more exposure to water, the softer and weaker the hemp will be. Water and washing are not recommended for the more complicated patterns.

Selling Hemp Jewelry - Be aware that many people are ignorant of the history and the possibilities of hemp products. To sell at an established business, ask permission from the manager or owner first, they like that! To sell wholesale to a business, cut the retail price by 50%. If selling by consignment, be careful and keep good books. In the process of sales, always smile and acknowledge the praise received. Failure to do so means asking for it twice. To acknowledge rejection, just smile and say thank you.

There are two basic knots used to make hemp jewelry, the Half Knot and the Half Hitch. Both knots are attractive and functional, which is one reason for their popularity in hemp jewelry projects. The other reason is that there are many variations of these two knots, which allows the knotter dozens of choices for pattern combinations. There are many *funky* knots which can be used with hemp; the best are in this guide.

In addition, this section gives knots for beginning and creating clasps or closures for hemp jewelry. An abbreviation is given in parentheses for each knot, and this abbreviation, followed by a number, is used to indicate which knots, and how many, are needed in the patterns at the end of the book.

Unless otherwise indicated all these knots are illustrated with four parallel cords. This is done by starting with four separate cords or two cords folded at the center.

Be aware that more than two knotters and any number of bead carriers can be used to create different effects in hemp jewelry. The basic procedure for each knot remains the same as when using four cords.

Beads are often threaded on the bead carriers between knots, or they can be strung on all four cords between knots. Actually, beads can be added to any of the cords, or any combination of cords, in any position which pleases the jewelry maker.

Lark's Head (LH) Mounting Knot

This knot is most often used for mounting cords on rods and rings. It can also be used for connecting cords. Cut the cord lengths as instructed for the pattern. Mount them one at a time, unless otherwise specified. Find the center of the cord and fold it in half. The Lark's Head knot may be done with the cord laying in front of or behind the rod or ring. Take the loop formed and wrap it over the top of the rod or ring. Bring both ends of the cord under the rod or ring and through the loop. Pull the loop snug (see **Figure 5**).

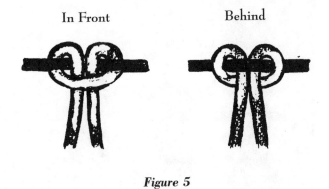

In Front   Behind

Figure 5

HALF KNOT AND ITS VARIATIONS

Half Knot (HK)

A Half Knot (HK) can be either right or left handed. If no direction is indicated in a pattern,

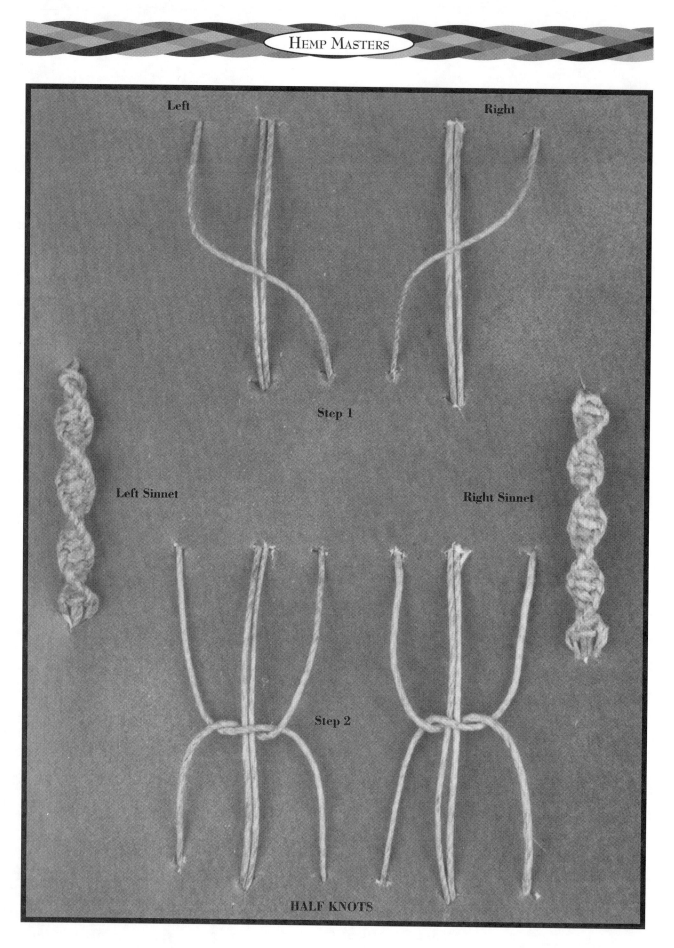

Left

Right

Step 1

Left Sinnet

Right Sinnet

Step 2

HALF KNOTS

use whichever is easiest.

Start a Right Half Knot (RHK), with the Right cord. Place it *over* the two center cords and *under* the Left cord. Let go of the Right cord and let it lay on the craft board (see **Step 1 Right** in photo). Take the Left cord *under* the two center cords and *up through* the loop made by the Right cord. Take care that the center two carriers remain in their original position. Pull the outside cords taut (see **Step 2 Right** in photo).

A Left Half Knot (LHK) is made using the same procedure as for a Right Half Knot, only the knot is started with the Left cord, followed by the Right cord (see **Steps 1 and 2 Left** in photo).

Half Knot Sinnet (HKS)

A Sinnet is a vertical chain or braid of repeated knots. Using Half Knots creates a spiral braid. To make a Half Knot Sinnet, repeat Steps 1 and 2 of the Half Knot several times. Use either all Right Half Knots or all Left Half Knots. After the first few knots are completed, the Sinnet will start to twist; the direction of the twist depends on the type of Half Knot used. Allow the piece to twist, turning the work as the knotting progresses (see **Half Knot** photograph). The number of knots it takes to complete a full twist trips heavy (depends) on the thickness of cord being used. The thinner the cord, the sooner it twists and the more twists per inch.

Overhand Knot (OK)

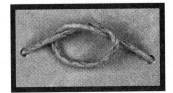

An Overhand Knot (OK) is actually just a Half Knot tied in a single cord. To tie this knot, simply make a loop near the end of the cord. Bring the end around the cord and through the loop, as shown in the photograph at right. Pull the end to tighten the knot; move the loop to position the knot as it is tightened.

Square Knot (SK)

The Square Knot consists of two Half Knots, one Right and one Left. Square Knots can be either right or left handed, depending on which type of Half Knot is tied first. When a Square Knot (SK) is indicated in a pattern without a right or left designation, use whichever is easiest to tie (see **Figure 6**).

Start a Right Square Knot (RSK) by completing Steps 1 and 2 for a Right Half Knot (see **Page 12**). With steps 1 and 2 done, and the four cords laying side by side on the board, take the Left cord and place it *over* the two center carriers and *under* the Right

| Step 3 | Step 4 | Finished Knot |

Figure 6

13

cord. Let go of the Left cord and let it lay on the board (see **Step 3**).

Take the Right cord *under* the two center carriers and up through the loop made by the Left cord (see **Step 4**). Pull the knotting ends taut.

Note that when square knotting is started with the right cord *over* the carriers, the completed knot will have the "hip" (the vertical cord that passes over the two horizontal cords in the knot) on the right side of the knot (see **Finished Knot**).

A Left Square Knot (LSK), illustrated in **Figure 7** is done in the same manner as a Right Square Knot, except that it be-gun on the left side with the Left cord. Start by completing Steps 1 and 2 for a Left Half Knot (see **Page 12**). Then take the Right cord and place it *over* the two center carriers and *under* the Left cord (see **Step 3**).

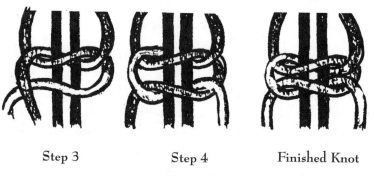

Step 3   Step 4   Finished Knot

Figure 7

Take the Left cord *under* the two center carriers and up through the loop made by the Right cord (see **Step 4**). Since the knotting is started with the Left cord *over* the carriers, the completed knot will have the "hip" on the left side of the knot (see **Finished Knot**).

Remember, in the Square Knot the same cord always crosses over the top of the carriers. When the four steps of the Square Knot are completed, the hip is on the same side as the cord used to start the knot.

Square Knot Sinnet (SKS)

To tie a Square Knot Sinnet, tie a series of Square Knots. Use all Right Square Knots to create a Right Square Knot Sinnet (RSKS) or all Left Square Knots to create a Left Square Knot Sinnet (LSKS). A Square Knot Sinnet lays flat. To deter-mine how many Square Knots have been tied, count the hips on the side from which the knotting was started. In **Figure 8** there are five hips and therefor five knots in each sinnet.

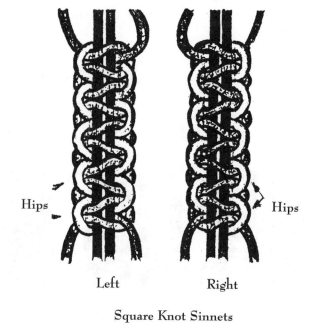

Hips   Hips

Left   Right

Square Knot Sinnets

Figure 8

Switch Knot (Switch)

The Switch Knot (Switch) is a hemp saver, it looks great in hemp jewelry, and it is easy to tie. To accomplish the Switch Knot, use two knotters and two carriers. Begin by putting the old knotters in front of and in between the two old carriers. Pull the old carriers away from each other and use them to tie a Square Knot around the old knotters, about 3/4" away from the last knot tied (see **Figure 9**). Now the old carriers are the new knotters and the old knotters are the new carriers.

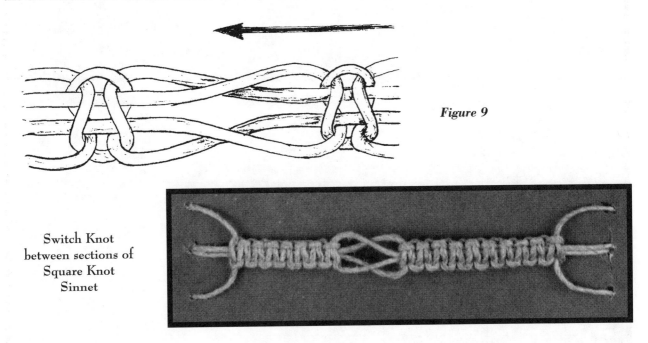

Figure 9

Switch Knot
between sections of
Square Knot
Sinnet

Alternating Square Knot (ASK)

Dig This! The Alternating Square Knot (ASK) is tied using multiple sets of four cords each. The result is an attractive basket-weave pattern. Start by pinning the sets of four parallel cords next to one another on the craft board. The example in the photograph on **Page 16** is exploded for clarity and uses three sets of cords, but any number may be used.

Tie a Square Knot in each set of cords; these knots should be side by side, in the same place on each set of cords (see **Row 1** in the photo).

Momentarily set aside the two outside cords from both the right and left hand knots. Use the two inside cords from the right hand knot and the two right hand cords from the middle knot to tie a Square Knot. Use the two left hand cords from the middle knot and the two inside cords from the left hand knot to tie a second Square Knot (see **Row 2**).

Place all the cords in their original positions, as at the beginning of Row 1. For Row 3 tie a Square Knot in the left, middle, and right hand sets of cords, as was done in Row 1 (see **Row 3**). Row 4 is tied in the same way as Row 2, and so the pattern is established. Continue in this manner for as long as desired. Try allowing more or less space between the rows for different effects.

Many variations of the Alternating Square Knot are possible. **Figure 10** shows two of these.

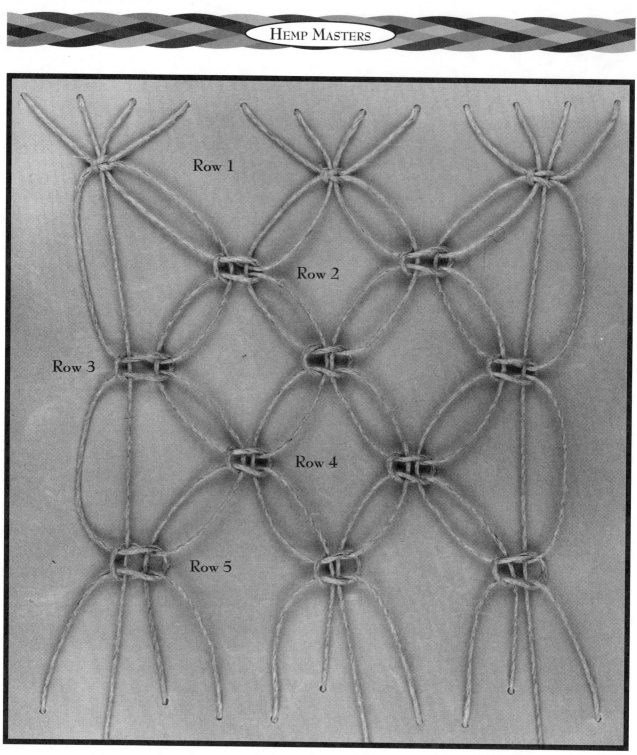

Row 1

Row 2

Row 3

Row 4

Row 5

Alternating Square Knot

The first drawing shows increasing the number of Alternating Square Knots in each row by one (in a multiple set of cords). The second drawing shows the opposite effect, created by decreasing the number of Alternating Square Knots in each row by one.

Increasing by one ASK per row   Decreasing by one ASK per row

Figure 10

Square Knot Button (SKB)

The Square Knot Button (SKB) is illustrated in the accompanying photographs (**Pages 17 & 18**). Begin by leaving a small space where the button will be, looping the knotters just enough to allow the carriers to be threaded through the loops (see **Step 1**). Then tie three Square Knots (or more if a larger button is desired).

Take the center carriers and pull them *forward* and *up* to the space above the first Square Knot tied. Thread the carriers through the small loops created in Step 1, from *front to back*, and bring them down the back side of the Square Knot series (see **Step 2**). If a button is desired on the back side of the piece, just thread the carriers from back to front during Step 2.

From the back side, pull the carriers until the last knot tied is snug against the small space created in Step 1 (see **Step 3**). The series of three Square Knots originally tied will now be rolled into a Button. Secure the Button in position by tying another Square Knot directly below the roll. Continue knotting if desired (see **Step 4**).

This is a great method for ending and provides a button for the beginning loop or

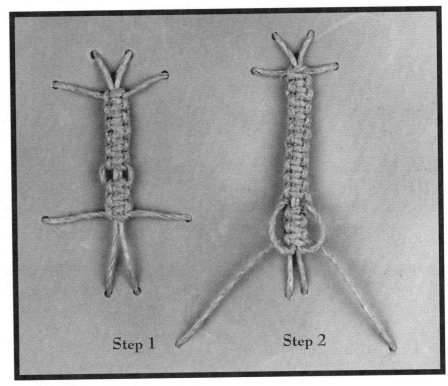

Step 1   Step 2

17

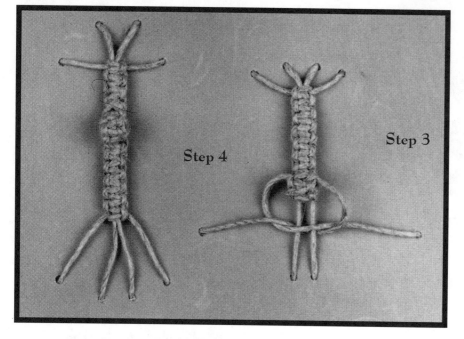

Step 4   Step 3

clasp to go around. The jewelry can be made adjustable in size by using more than one button.

Gathering Knot (GK)

A Gathering Knot (GK) is one mor more Square Knots tied around any number of cords to draw the group of cords together (see **Figure 11**). It may be used within a pattern or as a finishing step.

Figure 11

Butterfly or Picot Knot (BK)

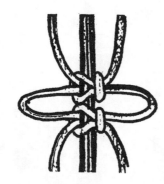

Step 1   Step 2

Figure 12

A Picot or Butterfly Knot (BK) is illustrated in **Figure 12**. It consists of two Square Knots, tied at a distance from one another and then drawn together. Tie the first Square Knot in the regular manner. Tie the second Square Knot a half inch (or more) below the first knot (see **Step 1**). Hold the carriers firmly and gently slide the second Square Knot up into position below the first knot (see **Step 2**). The knotter cord between the two knots forms loops as the knots slide together, resembling picot edging or butterfly wings.

HALF HITCH AND ITS VARIATIONS

The most important thing to remember about Half Hitch Knots is that the Anchor Cord or Bead Carrier must be placed *on top* of the Knotter Cord before the knotting begins. Also remember that the Anchor Cord must be held taut as the knot is tied or the Half Hitch will not form properly. Half Hitches can be tied with the Anchor Cord in a horizontal, vertical, or diagonal position. In patterns, the first two letters of the abbreviation will specify the position of the Anchor Cord.

Half Hitch (HH)

To make a Half Hitch (HH), start with the Bead Carrier or Anchor Cord (A) *on top* of the Knotter Cord (K). Hold the Anchor (A) taut and take the Knotter (K) *over* the Anchor. Then take the Knotter *under* the Anchor and *up* through the loop formed by the Knotter around the Anchor (so that the Knotter crosses over itself; see **Figure 13**).

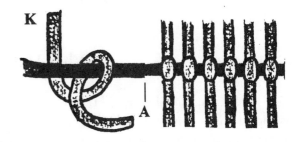

Figure 13

Note: A single Half Hitch tied in this manner only remains in position when there is tension on both ends of the Knotter, therefore it must be used in concert with other knots tied above and below the Half Hitch. To create a single Half Hitch which tightens on itself, go over the Anchor and *behind* the Knotter, before bringing the Knotter under the Anchor and through the loop (see photograph below).

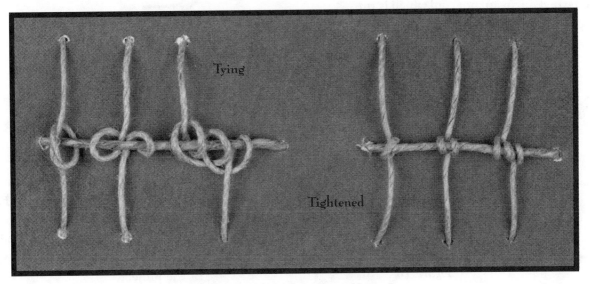

Single, Double and Triple Half Hitches

Double Half Hitch (DHH)

Two Half Hitches tied around a Carrier or Anchor Cord creates a Double Half Hitch (DHH; see

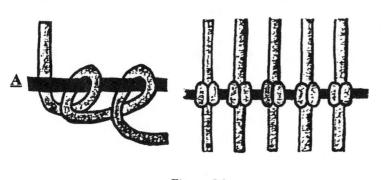

Figure 14

Figure 14). This is the most common way in which Half Hitches are tied. When completed, the Knotter cord will be positioned between the two Half Hitches and the knot is self-tightening. When attaching more than one knotter to an anchor cord, start with the first cord to the right of the anchor and continue left, one cord at a time, until all the desired cords have been tied around the anchor.

Other Half Hitch Variations

A Triple Half Hitch (THH), consisting of three consecutive Half Hitches (as the name implies) can also be tied around an anchoring cord, as shown in **Figure 15**.

Double or Triple Half Hitches can also be tied on an anchor cord which is running diagonally (see **Figure 16**).

A Half Hitch Sinnet (HHS) can

Figure 15

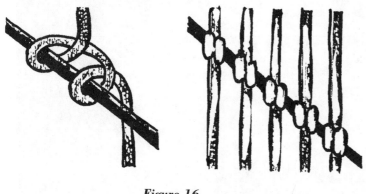

Figure 16

be created by tying a series of Half Hitches around a vertical anchoring cord. The sinnet can be either right-handed, with the loops on the right, or left-handed, with the loops on the left (see top left photograph on **Page 21**). More than one anchoring or carrier cord can be used depending on the number of cords being used in the project.

A different look can be achieved by switching the knotter and anchor cords between each Half Hitch tied. This is called an Alternating Half Hitch (AHH). This knot can be used to split a four cord piece into two halves, and is very attractive when tied with single cords and left spaced apart for an airy look, as in the top right photo-

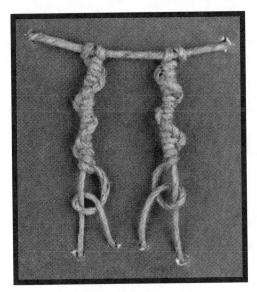

Half Hitch Sinnets
Mounted with
Lark's Head Knot

graph on this page.

The Alternating Half Hitch can also be tied with double cords, if the artist does not wish to split the work into two halves. Still another variation is created by using first one Knotter and then the other to tie Half Hitches around the Carriers.

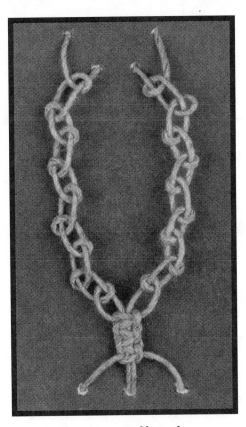

Alternating Half Hitch

STARTING AND FINISHING TECHNIQUES

Many types of closures are possible on necklaces, bracelets, and anklets. Most of them can be used on either end of the jewelry. Experiment a bit to find a favorite, as well as the type that best suits a particular project.

The easiest way to wear a piece of hemp jewelry is to leave some extra cord on the ends and tie them together with a Square Knot; however this should only be used if the jewelry will be worn until it falls off.

An Overhand Knot, tied using all four cords as a single cord, can be used to start or finish a piece of jewelry. A dab of glue will help hold it in place.

Square Knot

Overhand
Knot

Finishing or Starting Wrap (W)

This technique, illustrated in **Figure 17**, can be used to begin or end a piece of hemp jewelry. A three inch wrap requires a separate cord approximately two feet in length. To begin, lay the extra cord along side (parallel to) the cords to be wrapped. Form a loop at the top which goes to the bottom of the area to be

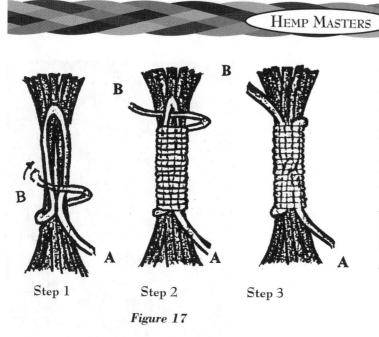

Step 1   Step 2   Step 3

Figure 17

wrapped (see **Step 1**). Take one end of the extra cord (marked **B**) and wrap from the **bottom up** as many times as desired, **pulling taut with each wrap**.

Take end "B" through the loop and hold it taught (see **Step 2**). Now pull on the other end of the extra cord (marked A) to bury the loop inside the wrap (see **Step 3**). **Do not** pull the end all the way through! Trim both ends close to the wrap. A bit of glue on the ends will help.

Loop Closures (LC)

These are a good starting point; first choose a bead or knot for the other end and then be sure to make a loop that the bead or knot can **just barely** pass through. For a Loop Closure at the start, begin with two looped cords, as indicated at the beginning of this Knotting Guide. Use the outside loop to tie a Square Knot (or any other knot desired in the pattern) around the inside loop, making sure that the end loop is the proper size (see **Figure 18**).

Figure 18

For a Loop Closure at the end of a knotted piece, tie off the knotters. Leave enough space for a loop along the carriers and then tie an Overhand Knot using both carriers as a single cord (see **Figure 19**). Cut off any remaining cords and glue all ending knots and cuts. Another way to create a loop on the end (or at the beginning for that matter) is to leave enough space along all four cords, then tie a few more knots before ending. Split the cords, two to each side, and pass the bead or knot between them.

For an adjustable loop closure at the start, make the loop larger and then slide a snug fitting bead over the loop and down to the knot (see **Figure 20**). When the loop is hooked around the opposite end of the piece, slide the bead along the loop to secure the closure.

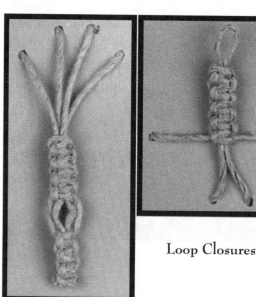

Loop Closures

Figure 19

Adjustable
Loop
Closure

Figure 20

Slide Loop Clasp (SLC)

This is a Hemp Master's original clasp which creates a beginning loop whose size is adjustable. It is the author's favorite way of beginning a piece. Start with two looped cords, one inside the other, as indicated at the beginning of this Knotting Guide. Use the outside loop to tie a Square Knot around the inside loop. Leave just enough of the inside loop at the top to fit over the width of the piece; remember, this slide is adjustable, so the loop does not have to fit over the knot or bead at the other end.

Continue tying four or five square knots, until there is enough to hold comfortably (about a thumb width; see photograph). Then Switch the knotters for the carriers (see **Figure 21**). About 3/4"

Figure 21

Slide Loop Clasp

below the last square knot, tie three Square Knots with the old carriers (around the old knotters). With the new carriers, tie a Half Knot snug against the last square knot. Continue tying Square Knots over the knot in the carriers. This will prevent knot slippage below the Switch.

To adjust the loop, grab the end of the loop with one hand and the section of square knots between the thumb and forefinger of the other hand. Slide the square knots down the carriers in the switch knot. The knotters will flare out to the sides. The square knots will stay at any point on the carriers, but the half hitch will ultimately stop their slide.

Bead Closures (BC)

Bead Closures are a good way to start or finish a pattern. Single or multiple beads can be strung on the starting loops before the knotting begins (see **Figure 22**). To create the same effect at the finish of a piece, cut one of the carriers and string the beads on the remaining carrier. Bring the end of the remaining carrier back up to the last knot tied, forming a loop. Secure the ends of the loop by tying 3 or 4 knots with the knotters. Cut and glue the ends of the knotters to finish the piece.

For a bead catch at the end of a pattern, tie off and glue the knotters.

Figure 22

Then string the chosen bead(s) on the carriers and tie an Overhand Knot using both carriers as a single cord. Glue the knot (see **Figure** 23 and left photograph). Another way to create a bead closure on either end is to "tie in" the bead (s). First string a bead on the carriers. Then bring the knotters around

Figure 23

either side of the bead and tie another knot. The photograph on the right shows two beads strung on the starting loop and a "tied in" bead which could be on either end.

Many variations are possible. For instance, crossing the carriers through a bead will change the way it lies (see **Figure** 24 and

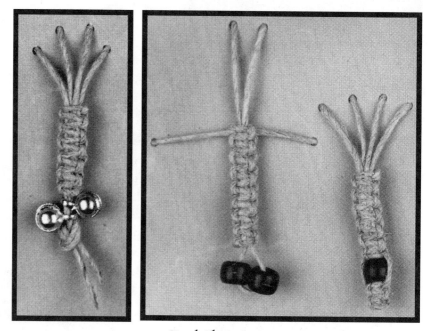

Bead Closures

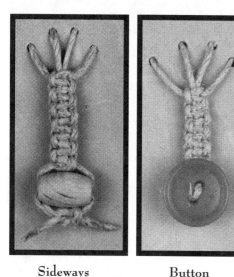

Sideways
Bead

Button
End

Figure 24

photograph on left). Or, if the beads are big enough, all four cords can be threaded through and then tied off. Buttons can also be used. Simply thread one or two carriers through the holes (two or four holes works best) and tie a knot on the back side of the button. Cut the carriers and glue the cut ends. Tie a knot over the cut ends, cut the knotters and apply more glue (see photograph on the right).

Ending Knots

To make a T-shaped end to fit through a loop, cut and glue the carriers, then take each knotter individually and tie an Overhand Knot on either side of the piece, snug against the sides of the last Square Knot (glue helps).

Adjustable sizing can be created by tying Overhand Knots to the side, without cutting the carriers, before the end of the piece. An end with T hooks such as these is called a Knuckled End.

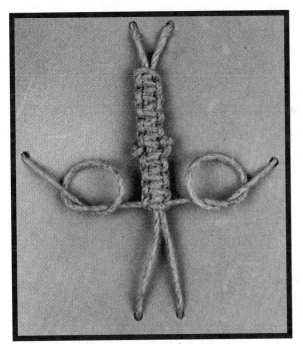

Tying *T* Hooks

To create a "bump" for a loop to catch on, tie a Half Knot on *top* of the last knot tied, then tie a few more knots before ending the piece. This technique can also be used to create adjustable sizing by tying Half Knots on top of the pattern at spaced intervals from the end of the piece. An end with "bumps" such as these is called a Ripple End.

A simple way to end a piece of hemp jewelry is to cut the two middle carriers and put some glue on the cut ends. With the knotters, tie a tight Half Knot over the glued ends, then cut off the knotters. This ending technique is shown on the knuckled

Knuckled End

end pictured above.

To make a larger ending knot (End), which can be fastened through a loop, tie the knotters back over the end of the pattern. Use the same knots as were used on the end of the pattern. Glue the last knot and cut off the knotters. This ending technique is one of the author's favorites and is pictured on many of the project pieces in the next section.

KNOTS FOR MASTERS

Pretzel Knot (PK)

This knot is very seldom used by the new generation of hemp artists, but it may still look familiar (deja vu). The Pretzel Knot (PK) has been used in the U.S. as a decoration on military uniforms from the 1800's to the present day. It is also seen in the decorative weaves of modern baskets, macramé, door mats and rugs. And of course, Hemp Masters artists know how to use this knot in their hip, hemp jewelry.

The Pretzel Knot has a *style*, a look, an aura if you will, of its own. Many say it is full, complete, or warm; that it makes a person feel welcome or safe.

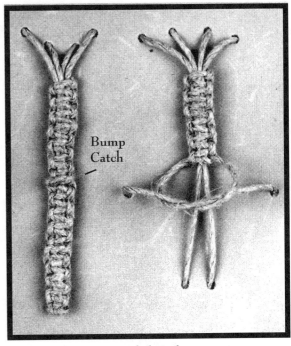

Rippled End

Others say it looks Celtic, strong, ancient or even medieval. Whatever type of jive they put on this *trippy hippie hemp twist*, it shows unique balance, for its grace is complete and defined. It is admired by all who see it.

Remember the following when tying a Pretzel Knot: This knot can be achieved with two, four, or any multiple of two cords. The photographs that illustrate these instructions show the structure of the Pretzel Knot with four cords, but feel free to use six or eight cords. **Be sure all cords remain parallel** to each other, through the entire knot, as the pretzel knot is carefully tightened. If more than one of these knots is tied in sequence, **each subsequent knot must start on the opposite side from the previous knot tied.** This will prevent the pattern from twisting.

The example which accompanies these instructions begins at the end of a square knot sequence, but it could be at the end of any sequence of knots. To start, split the cords at the end of the knotting sequence in half. In the example, the left knotter and left carrier are pulled to the left (the "left set") and the right knotter and right carrier are pulled to the right (the "right set").

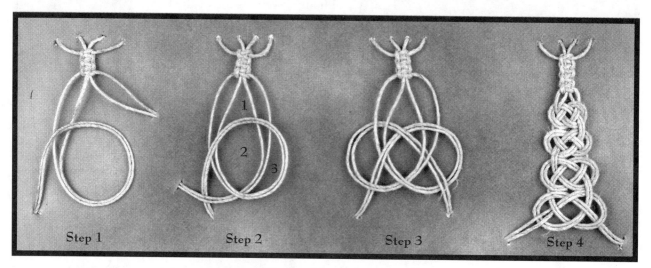

Tying a Pretzel Knot

Begin this knot on either side, just be sure that the starting loop always crosses over or in front of itself (just once) and that the loop always faces the opposite set of cords (or the center of the piece, whichever is easier to visualize). Make a loop with one of the sets of cords; in the example the left set has been used to make the starting loop (see **Step 1**). Notice that the left set of cords crosses over (in front of) itself and that the loop is facing the right set of cords (towards the center). Hold the loop in the left thumb and forefinger, at the point where the left set of cords crosses itself.

With the right hand, bring the right set of cords **behind the starting loop**, so that it divides the loop in half, from top to bottom (see **Step 2**). With the right thumb and forefinger, hold the knot at the top of the loop, where the right set of cords crosses behind the top of the loop. Let go of the pattern with the left hand.

Follow the right set of cords down to about five or six inches below the big loop and, with the left hand, fold this set of cords to create a point or hemp needle. This will make it easier to weave the right set of cords through the loops created thus far. Make sure the right set of cords is over, or in front of, the left set of cords that lay below the big loop they were used to create (see **Step 2**). (*Hey man, don't freak out*, get a grip and read it again.)

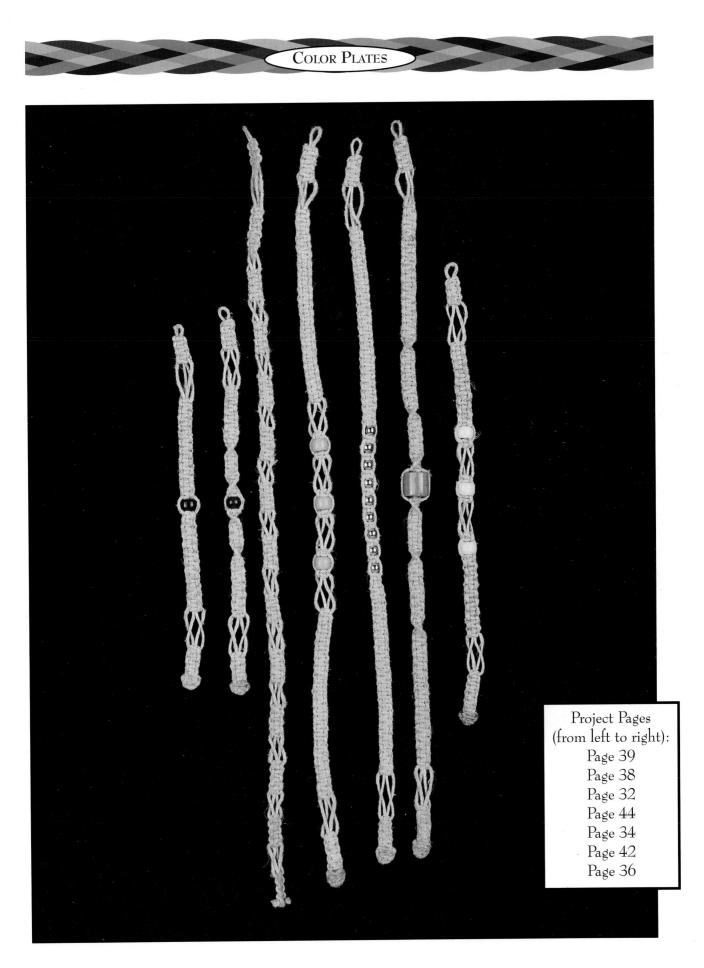

Project Pages
(from left to right):
Page 39
Page 38
Page 32
Page 44
Page 34
Page 42
Page 36

Plate I

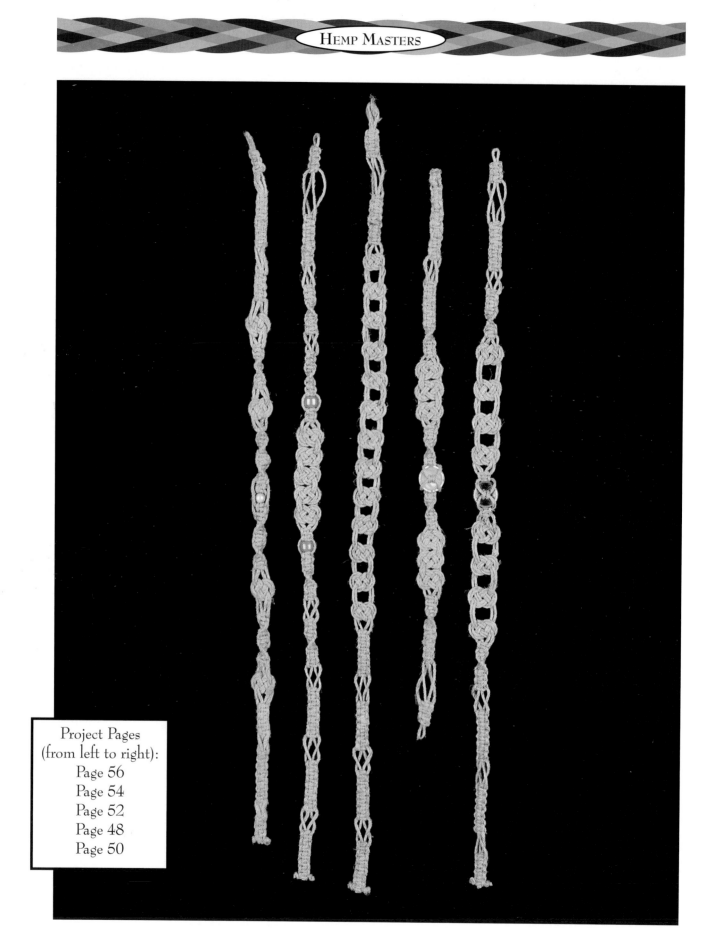

Project Pages
(from left to right):
Page 56
Page 54
Page 52
Page 48
Page 50

Plate II

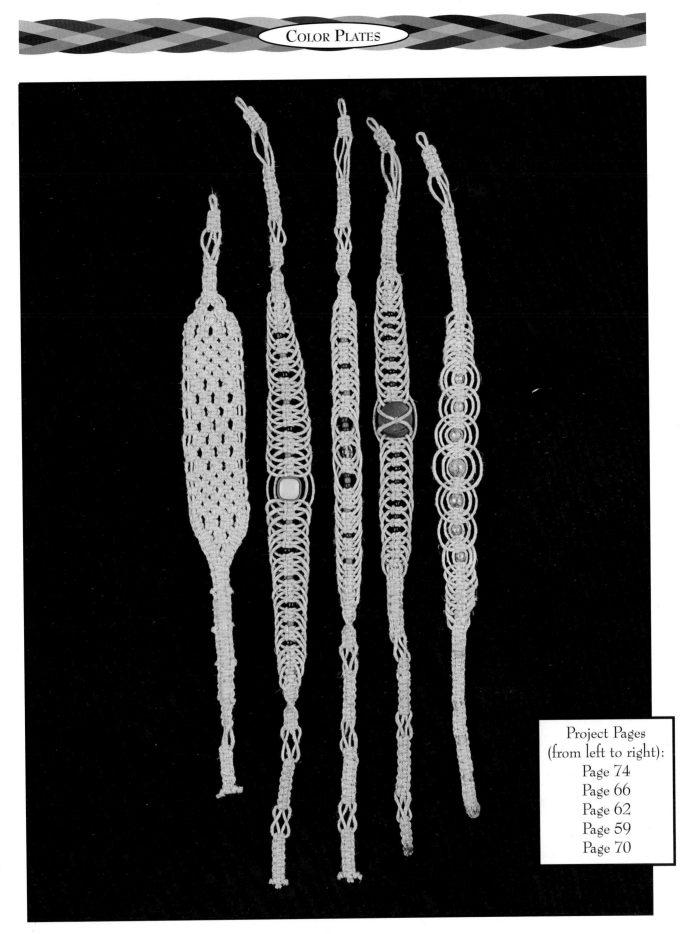

Project Pages
(from left to right):
Page 74
Page 66
Page 62
Page 59
Page 70

Plate III

Project Pages
(from left to right):
Page 78
Page 84
Page 80

Plate IV

Notice that there are now three half loops. Above the big loop is a half loop created by the right and left sets of cord (**Hole 1**), and the big loop is divided into two half loops, top (**Hole 2**) and bottom (**Hole 3**).

With the hemp needle in the left hand, place the right set of cords behind the top left side of the knot and up through the top half loop (Hole 1). Then go over the top of the big loop and down into the middle half loop (Hole 2). Continue behind the right set of cords that divides the big loop in half and bring the hemp needle up through the bottom half loop (Hole 3). This is shown completed, but not tightened, in **Step 3**. The weaving is now finished, so release the fold of the hemp needle and pull the right set of cords out of Hole 3.

Make sure all cords in each set are parallel to one another before pulling the knot tight. Pull on the two sets of cords (left and right) and the Pretzel knot will begin to shrink and take shape. To position the Pretzel Knot at the desired point in the pattern, pull on the bottom right and left loops of the knot. Repeat this process until the knot is in the proper position and is the size desired (see **Step 4**).

Finished Hemp Jewelry Pieces
Containing Pretzel Knots

Phish Bone (PB)

This knotting sequence takes a while to form its pattern. It can be tied with any number of knotter sets, but only one, lighter weight carrier is used. The more knotters, the wider the piece and the longer it takes for the pattern to form.

Setting up the cords for this knotting sequence is a very flexible process and depends on the vision of the artist. The carrier can be included at the beginning of the piece or it can be added at any point along the way. All the knotters can be added at once, or they can be added at a distance from one another. While tying this pattern, keep a tight twist in all the cords to maintain the definition of the Phish Bone pattern. Beads can be added at any point in the Phish Bone; they have been used in this example to show how much panache they can add to a piece of hemp jewelry.

The photographs that illustrate these instructions start with a standard two knotter, two carrier setup, in which a Square Knot Sinnet has been tied. To set up for the Phish Bone, this is converted to three sets of knotters (all 45 lb. test) and a single 20 lb. test carrier. This example set up is just one of many possibilities.

    The first step is the addition of the 20 lb. carrier. Adding a carrier is illustrated in the Ancient Hippie Secrets and Helpful Hints section (see **Page 8**). Cut the new carrier to the same length as the original carriers. Place the new cord parallel (next) to the existing carriers. Pull the top of the new cord about two inches above the last knot tied. With the knotters, tie a Half Knot around all three carriers. Take the two inch section of the new cord and fold it down over the Half Knot just tied, so it lies parallel to the other carriers. With the knotters, tie another Half Knot (the second half of a Square Knot) around the four carriers (see **Step 1**). Tie three more Square Knots over all four carriers to hold everything securely in place (see **Step 2**). Cut the short (2") carrier if necessary.

    Next add a new 45 lb. cord. Simply place the center of the new cord across (at right angles to) the carriers. Place the original knotters over the new cord and tie two Square Knots (see Adding Extra Cords, **Page 8**). Leave the new cord sticking out either side of the piece. To finish the set up, leave the original knotters to either side and tie two Square Knots with the two 45 lb. carriers. Leave these to either side as a third set of knotters (see **Step 3**).

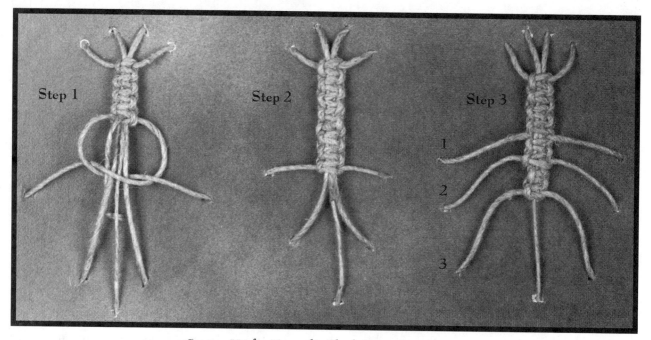

Setting Up for Tying the Phish Bone Pattern

    There are now three sets of knotters (1, 2, & 3 from top to bottom) and one bead carrier. Place the #1 knotters in front of the other two sets and tie a Square Knot tight, just below the #3 knotters. Leave the #1 knotters to either side. Now place the #2 knotters in front of the #3 and the #1 knotters (beside the knot just tied) and tie a Square Knot tight, just below the #1 knotters. Leave the #2 knotters to either side of the piece. Place the #3 knotters in front of the #1 and #2 knotters and tie a Square Knot tight. Leave the #3 knotters to either side. **Be very aware of the size of the loops made by the knotters**; they should be either the same size throughout the pattern or each loop can gradually increase in size to the center point and then gradually decrease on the far side of the center point.

    String a bead on the carrier and slide it up to the knot just tied. Place the #1 knotters in front of the #2 and #3 knotters and tie a Square Knot just below the bead (see **Step 4**). Slide everything snug and adjust the size of the knotter loops before the final tightening of the knot below the bead.

Repeat this knotting sequence until the center point of the choker, bracelet, anklet, whatever, is reached (the second knot has already been tied with the #1 knotters, so start with the #2 knotters). String a center bead onto the bead carrier and snug it against the last knot tied

Hey man, remember to check the size of the piece against your body to establish where the center bead should be placed. This technique also helps in deciding on other bead placements when creating an original Fish Bone pattern.

To create the center point of the pattern, take the set of knotters **closest** to the center bead, go around either side of the center bead and tie a Square Knot tight against the other side of the center bead. Flip or rotate the entire piece over or face down (depending on your state of mind); this is labeled a

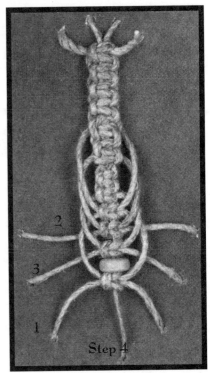

Tying the Phish
Bone Pattern

Flip on the patterns. Using the next set of knotters above the center bead, pull them around the center bead and in front of the knotters just tied. Check the size of the loop and tie another tight Square Knot. Repeat this process with the third set of knotters. To complete the center point of the Phish Bone Pattern, bring the knotters closest to the center bead **in front of** the other two sets of knotters and tie a Square Knot.

Begin tying the second half of the Phish Bone pattern using the same knotting sequence given for the first half. The knotting and beading sequence from the center point to the end of the Phish Bone pattern must exactly match the knotting and beading sequence from the center point to the beginning of the Phish Bone pattern to give the jewelry balance. While it is not absolutely necessary to tie a center point, it certainly adds to the overall appearance of the pattern and your *bank*.

After the Phish Bone knotting sequence is completed, the extra cords must be secured and then cut so that the Square Knot Sinnet from the beginning of the piece can be repeated. To do this place all the cords, except the last set of knotters used, next (parallel) to the bead carrier. Use the

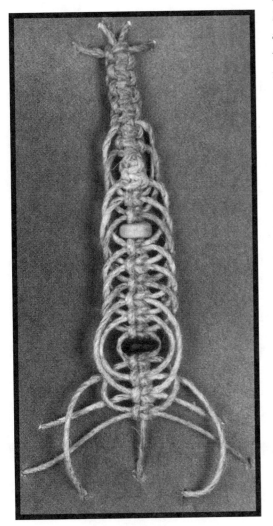

Center of Phish Bone
Pattern Completed

29

remaining set of knotters to tie a Gathering Knot (use three square knots) tightly around all of the cords gathered in the center. Leave two of the heavier cords in the center to serve as carriers and cut the remaining center cords about 1/4 inch from the last knot tied. Apply some glue to the cut ends so that they will not slip. Tie a Square Knot Sinnet until this groovy creation is the desired length. This is one of the *coolest*, *bank maker*, patterns ever!

Hempen!

I know some knots with Optic Manipulation
So Dig on this, with Heavy Meditation.

They're Groovy, they're Cool,
They're one of a kind,
The coolest of knots,
They blow everyone's mind.

Be strong on Imagination
And soon it will be.
Don't rush to frustration
For it's beauty you'll see.

It's a real "Attention Taker"
They say it's the WOW
A "Heavy Bank Maker"
In my Knotting Know How.

As you create a Pretzel
Take your time, read it twice.
You must pay close attention
To create something nice!

If you feel your creations
Have Hemp Master ware.
Just mail us a picture
You can trust we'll be fair.

PATTERNS

Quick Knotting Guide Reference and Abbreviations

Most of the pictured examples for patterns in this section were created with 45 lb. test (approx.

2 mm) cord. The knotting was pulled rather tight. The patterns are designed to use different knotting combinations that have proven to be popular in selling to hemp enthusiasts. Don't be afraid to use these patterns as a take off point for experimenting with different knotting styles and combinations. Use imagination and enjoy the outcome of your artistic creation.

The patterns give an estimated length for the jewelry, but knotting combinations may need to be added or subtracted to achieve the length desired. This is especially true if the knotting being done is looser or tighter than the example. The lengths given for the hemp are deliberately on the long side.

To determine the length of knotting to add or subtract, measure a piece of jewelry which is already the desired length and compare it to the length given for the sample piece. Another method is to wrap a piece of hemp around the body part which will be enhanced with the jewelry; compare this length to the length given for the pattern being used and adjust accordingly. Or just measure the piece against the target body part as it is being created. Plan your creation, keeping the center point of the pattern in mind and compare the project to the body part being decorated and the pattern being used.

Wherever possible, add or subtract knots on **both** sides of the middle to maintain the symmetry of the design. The goal, once the center of the pattern is reached, is to reverse the knotting sequence, knot for knot, so that the pattern from the center to either end of the piece is exactly the same. However, if the plan doesn't work as anticipated, don't panic. Just shorten or lengthen the finishing end as needed; in the unlikely event that anyone notices, just tell them it's supposed to be that way! (What a concept!)

SIMPLE SWITCH KNOT CHOKER

Sample length: 16 inches
Materials: (2) 6 foot lengths of 45 lb. test hemp

This piece can also be wrapped twice around the wrist and worn as a bracelet or anklet. It also makes an attractive headband. Adjust the length, and wear it on practically any body part.

Step 1:   Fold the two cords in half and tie a Slide Loop Clasp (shown as tied, with loop closed).

Step 2:   Tie a Square Knot Sinnet which is 5 Square Knots long (a flat sinnet).

Step 3:   Tie a short Switch Knot, which is about a half inch long.

Step 4:   Tie 4 Square Knots.

Step 5:   Tie another short Switch Knot.

Step 6:   Tie 4 Square Knots.

Step 7:   This pattern does not have a center, so there is no need to worry about the two halves matching. Simply alternate 1 Switch Knot and 4 Square Knots until the desired length is reached.

Step 8:   Cut the carriers, put some glue on the cut ends, and tie one more Square Knot over the glued ends.

Step 9:   With each of the knotters, tie an Overhand Knot to the side of the piece. Add glue to each of these knots. This creates a "T hook" to catch in the Slide Loop Clasp.

SIMPLE SWITCH KNOT CHOKER

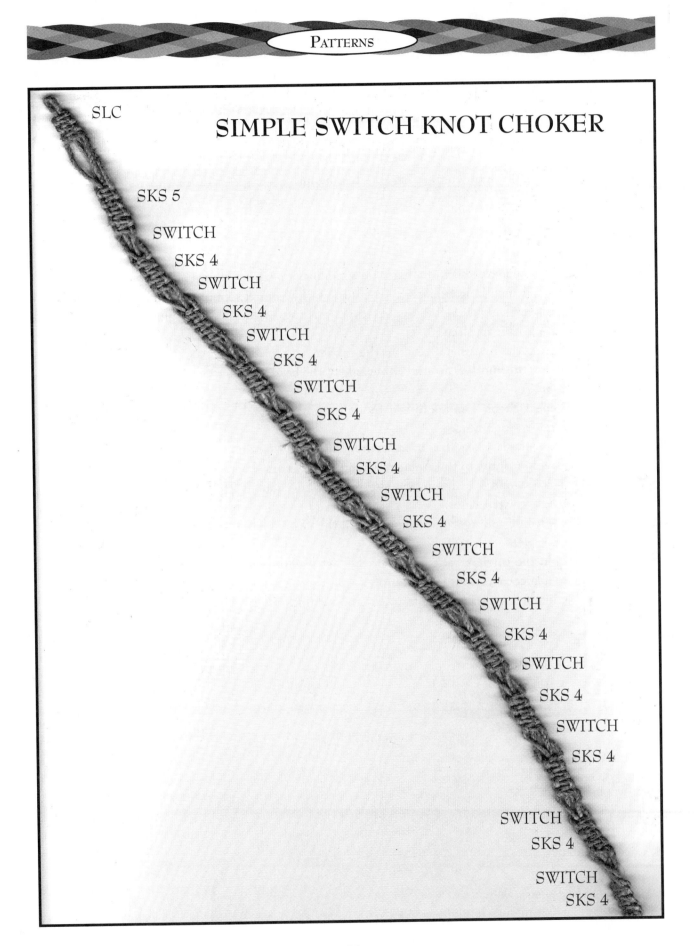

SLC

SKS 5

SWITCH

SKS 4

SWITCH

SKS 4

SWITCH

SKS 4

SWITCH

SKS 4

SWITCH

SKS 4

SWITCH

SKS 4

SWITCH

SKS 4

SWITCH

SKS 4

SWITCH

SKS 4

SWITCH

SKS 4

SWITCH

SKS 4

SWITCH

SKS 4

SWITCH

SKS 4

SWITCH

SKS 4

SQUARE KNOT CHOKER WITH METAL BEADS

Sample length: 15 inches

Materials: (2) 7 foot lengths of 45 lb. test hemp
                   (9) 3/16" metal beads (nickel or silver in the example)

This piece can be shortened on both sides of the beads and worn as an anklet or bracelet.

Step 1:   Fold the two cords in half and tie a Slide Loop Clasp (shown as tied, with loop closed).
Step 2:   Tie a 4 inch long Square Knot Sinnet (a flat sinnet; about 25 knots). Cut one of the bead carriers just before the last Square Knot is tied. A dab of glue on the cut end will help hold it in place.
Step 3:   String the first metal bead on the carrier.
Step 4:   Bring the knotters around either side of the bead and tie a Square Knot snug against the bead.
Step 5:   Repeat Steps 3 and 4 to add the remaining 8 metal beads, with a single Square Knot between each bead.
Step 6:   After the last bead has been secured, add a new (second) carrier. (See **Page 8** - Carriers Are Too Short in the Ancient Hippie Secrets and Helpful Hints section.)
Step 7:   Tie a second 4 inch long Square Knot Sinnet (about 25 knots).
Step 8:   Add a Switch Knot.
Step 9:   Continue adding Square Knots until the desired length is reached (6 in the example).
Step 10:   Fold the carriers back (up) over the last square knot and tie 2 Square Knots over the folded carriers and the last knot.
Step 11:   Cut the carriers and the knotters and saturate the cut ends with glue.

SQUARE KNOT CHOKER WITH METAL BEADS

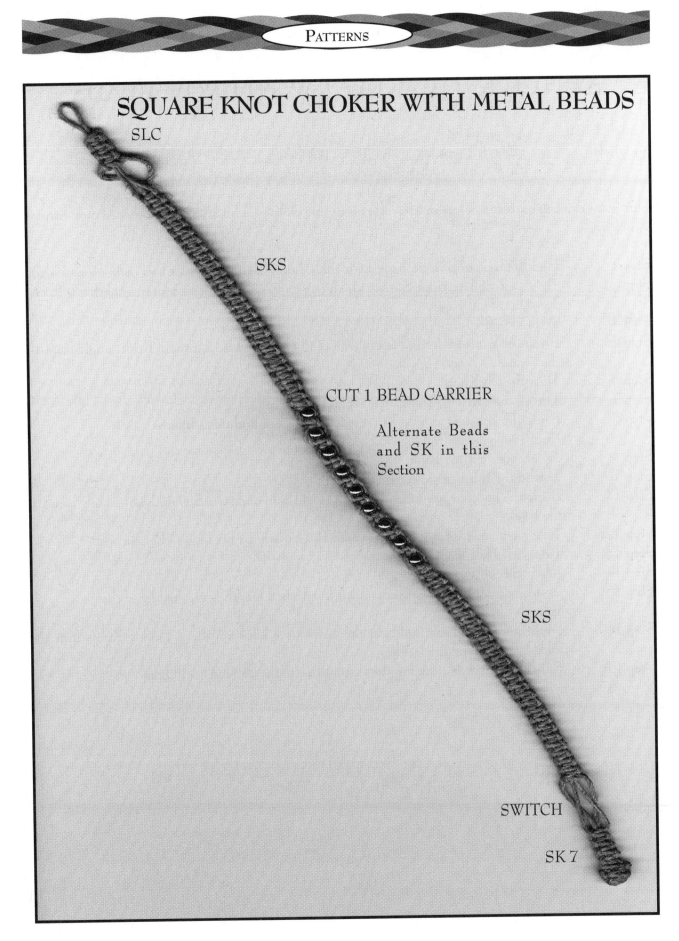

SLC

SKS

CUT 1 BEAD CARRIER

Alternate Beads
and SK in this
Section

SKS

SWITCH

SK 7

SWITCH KNOT ANKLET WITH CROW BEADS

Sample length: 9.5 inches
Materials: (2) 4 1/2 foot lengths of 45 lb. test hemp
                     (3) 9 mm crow beads (pearly white in the example)

This piece can be shortened before or after the beads and worn as a bracelet or lengthened and worn as a choker.

Step 1:   Fold the two cords in half and tie a Slide Loop Clasp (shown as tied, with loop closed).
Step 2:   Tie a Left Square Knot Sinnet which is 9 Square Knots long (a flat sinnet).
Step 3:   Tie a Left Half Knot.
Step 4:   String the first crow bead on the carriers.
Step 5:   Bring the knotters around either side of the bead and tie a Square Knot snugly beneath the bead.
Step 6:   Tie a Switch Knot.
Step 7:   Thread the carriers through the second crow bead.
Step 8:   Bring the knotters around either side of the bead and tie a Square Knot snugly beneath the bead.
Step 9:   Tie another Switch Knot.
Step 10:   Thread the third crow bead on the carriers.
Step 11:   Bring the knotters around either side of the bead and tie a Square Knot snugly beneath the bead.
Step 12:   Tie a Left Square Knot Sinnet which is 9 Square Knots long (a flat sinnet).
Step 13:   Tie a Left Half Knot.
Step 14:   Add a Switch Knot.
Step 15:   Continue adding Square Knots until the desired length is reached (7 in the example).
Step 16:   Fold the carriers back (up) over the last square knot and tie 2 Square Knots over the folded carriers and the last knot.
Step 17:   Cut the carriers and the knotters and saturate the cut ends with glue.

SWITCH KNOT ANKLET WITH CROW BEADS

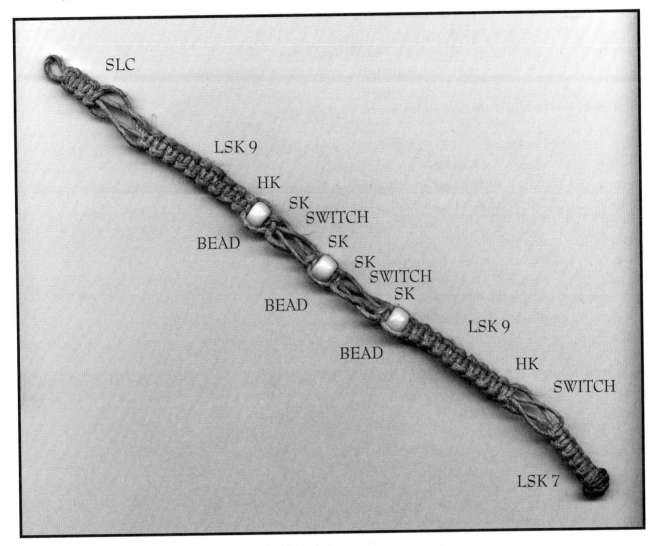

SLC

LSK 9

HK

SK

SWITCH

BEAD

SK

SK

SWITCH

SK

BEAD

LSK 9

BEAD

HK

SWITCH

LSK 7

HALF KNOT TWIST BRACELET

Sample length: 8.0 inches
Materials: (2) 4 foot lengths of 45 lb. test hemp
(1) 9 mm crow bead (transparent red in the example)

This piece can be lengthened on both sides of the twist and worn as an anklet.

Step 1: Fold the two cords in half and tie a Slide Loop Clasp (shown with loop completely open, i.e. with switch knot collapsed).
Step 2: Tie a 1 inch long Left Square Knot Sinnet (a flat sinnet; about 7 knots).
Step 3: Tie 5 Right Half Knots (a spiral sinnet).
Step 4: Tie 2 1/2 Left Square Knots, that is, 2 Left Square Knots followed by a Left Half Knot.
Step 5: Add 5 Right Half Knots.
Step 6: Run the carriers through the bead.
Step 7: Bring the knotters around either side of the bead and tie a Left Half Knot snug against the bead.
Step 8: Tie 4 more Left Half Knots.
Step 9: Tie 2 1/2 Left Square Knots, that is, 2 Left Square Knots followed by a Left Half Knot.
Step 10: Add 5 more Left Half Knots.
Step 11: Tie a 1 inch long Left Square Knot Sinnet (a flat sinnet; about 7 knots).
Step 12: Tie a Switch Knot.
Step 13: Continue adding Square Knots until the desired length is reached (5 in the example).
Step 14: Fold the carriers back (up) over the last square knot and tie a Square Knot over the folded carriers and the last knot.
Step 15: Cut the carriers and the knotters and saturate the cut ends with glue.

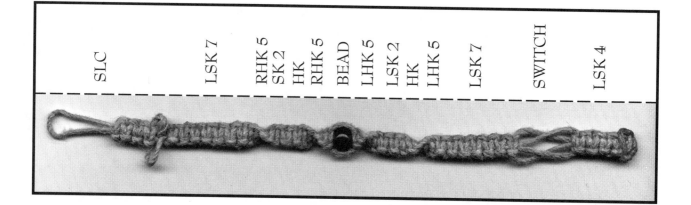

SLC | LSK 7 | RHK 5 | SK 2 | HK | RHK 5 | BEAD | LHK 5 | LSK 2 | HK | LHK 5 | LSK 7 | SWITCH | LSK 4

SQUARE KNOT BRACELET WITH CROW BEAD

Sample length: 7.5 inches
Materials: (2) 4 foot lengths of 45 lb. test hemp
                   (1) 9 mm crow bead (transparent dark brown in the example)

This piece can be lengthened on both sides of the bead and worn as an anklet.

Step 1:   Fold the two cords in half and tie a Slide Loop Clasp (shown with loop completely open, i.e. with switch knot collapsed).

Step 2:   Tie a 2 inch long Left Square Knot Sinnet (a flat sinnet; about 12 knots).

Step 3:   String the crow bead on the carriers.

Step 4:   Bring the knotters around either side of the bead and tie a Left Square Knot snug against the bead.

Step 5:   Tie a 2 inch long Left Square Knot Sinnet (a flat sinnet; about 12 knots).

Step 6:   Tie a Switch Knot.

Step 7:   Continue adding Square Knots until the desired length is reached (4 in the example).

Step 8:   Fold the carriers back (up) over the last square knot and tie 2 Square Knots over the folded carriers and the last knot.

Step 9:   Cut the carriers and the knotters and saturate the cut ends with glue.

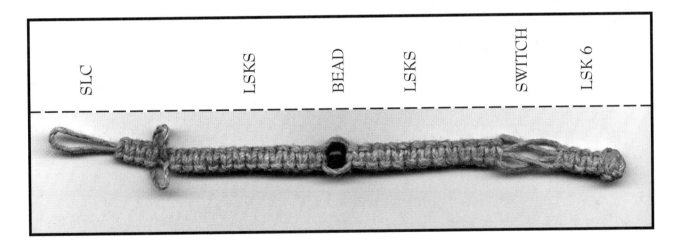

SLC   LSKS   BEAD   LSKS   SWITCH   LSK 6

PRETZEL KNOT ANKLET

Sample length: 12.75 inches
Materials: (2) 6 foot lengths of 45 lb. test hemp

This piece can be shortened after the Pretzel Knots and worn as a bracelet or lengthened before the Pretzel Knots and worn as a choker.

Step 1:   Fold the two cords in half and tie a Slide Loop Clasp (shown as tied, with loop closed).

Step 2:   Tie 2 Left Square Knots. Separate the four cords into two sets, pulling the left knotter and carrier to the left and the right knotter and carrier to the right. Leave about 1/8" in each set of cords and then tie a slightly open Pretzel Knot.

Step 3:   Leave another 1/8" in the cords and tie a second slightly open pretzel Knot. Start this Pretzel Knot on the opposite side from the first one.

Step 4:   Continue leaving 1/8" spaces and tying Pretzel Knots until a total of 7 Pretzel Knots have been tied. Note: an Attractive Variation of this pattern is to increase the openness of each Pretzel Knot until the fourth (center) knot is reached, then decrease the openness of each remaining Pretzel Knot to the same degree.

Step 5:   Leave a 1/8" space after the last Pretzel Knot, then tie 7 Left Square Knots.

Step 6:   Tie a Switch Knot.

Step 7:   Tie 7 Left Square Knots.

Step 8:   Add another Switch Knot.

Step 9:   Tie 4 Left Square Knots.

Step 10:   Tie a Switch Knot.

Step 11:   Add 7 Left Square Knots.

Step 12:   Cut the carriers and glue the ends (saturate the hemp fibers). Take the knotters individually and tie a Half Hitch to either side of the end of the piece. Glue these knots as well.

PRETZEL KNOT ANKLET

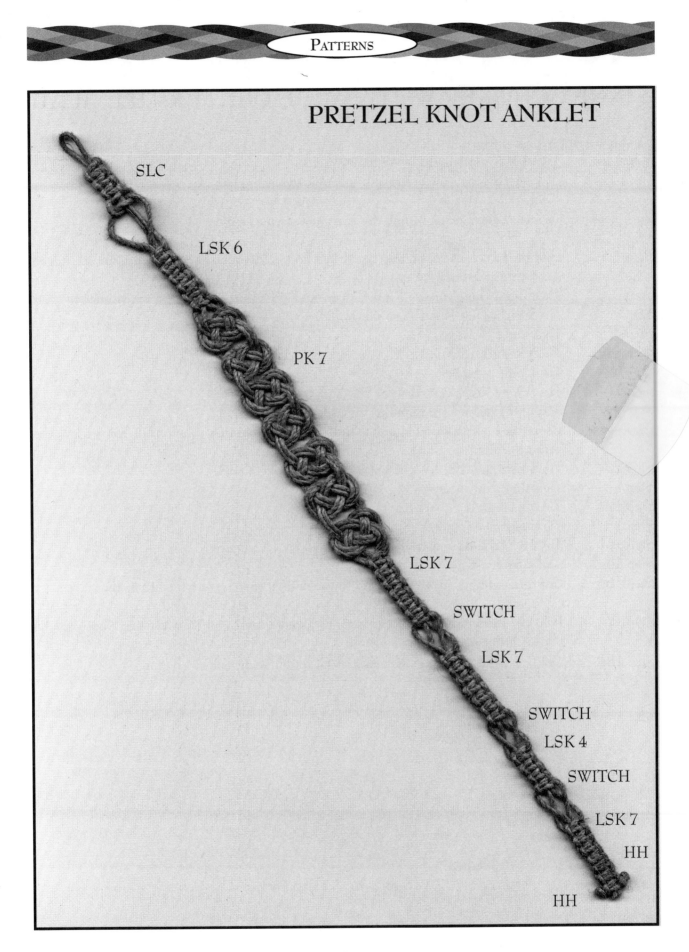

SLC

LSK 6

PK 7

LSK 7

SWITCH

LSK 7

SWITCH

LSK 4

SWITCH

LSK 7

HH

HH

SINNET CHOKER WITH LARGE CENTER TILE BEAD

Sample length: 15 inches
Materials: (2) 7 foot lengths of 45 lb. test hemp
                   (1) 1/2 inch tile bead (kelly green in the example)

This piece can also be wrapped twice around the wrist and worn as a bracelet.

Step 1:   Fold the two cords in half and tie a Slide Loop Clasp (shown with loop completely open, i.e. with switch knot collapsed).
Step 2:   Tie a 2 3/4" long Square Knot Sinnet (about 16 Square Knots; a flat sinnet).
Step 3:   Tie 5 Right Half Knots (a spiral sinnet).
Step 4:   Tie 6 more Square Knots.
Step 5:   Tie 5 Right Half Knots.
Step 6:   Tie 4 Square Knots.
Step 7:   Tie 5 more Right Half Knots.
Step 8:   String the carriers through the large tile bead.
Step 9:   Bring the knotters around either side of the bead and tie a Left Half Knot snugly beneath the bead.
Step 10:   Tie 4 Left Half Knots, followed by 4 Left Square Knots.
Step 11:   Tie 5 Left Half Knots.
Step 12:   Tie 6 more Left Square Knots.
Step 13:   Tie 5 Left Half Knots.
Step 14:   Tie a 2 3/4" long Square Knot Sinnet (about 16 Square Knots).
Step 15:   Add a Switch Knot.
Step 16:   Continue adding Square Knots until the desired length is reached (5 in the example).
Step 17:   Fold the carriers back (up) over the last square knot and tie a Square Knot over the folded carriers and the last knot.
Step 18:   Cut the carriers and the knotters and saturate the cut ends with glue.

SINNET CHOKER WITH LARGE CENTER TILE BEAD

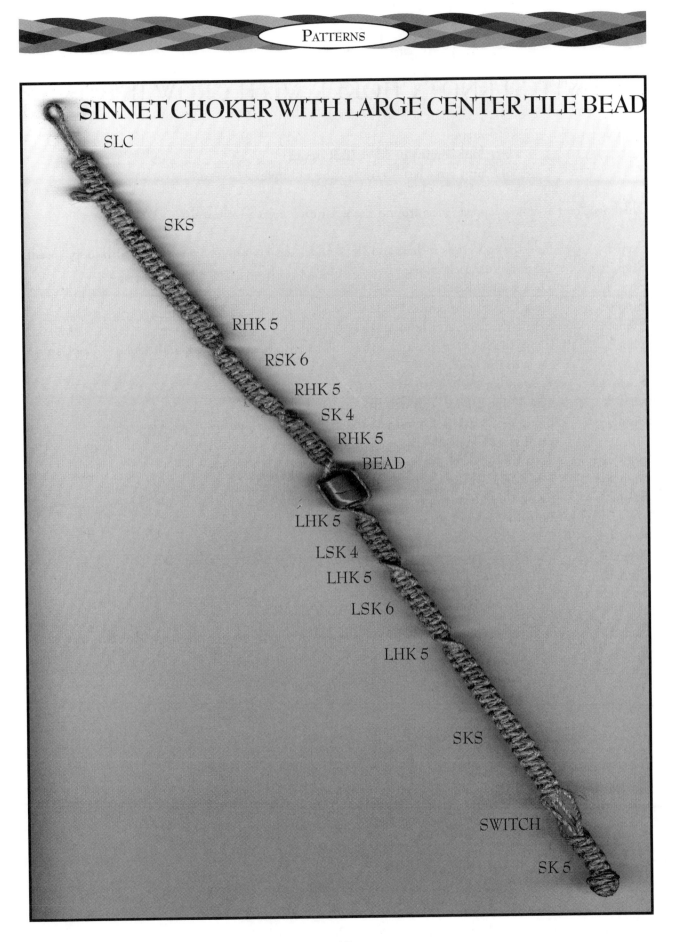

SLC

SKS

RHK 5

RSK 6

RHK 5

SK 4

RHK 5

BEAD

LHK 5

LSK 4

LHK 5

LSK 6

LHK 5

SKS

SWITCH

SK 5

SWITCH KNOT CHOKER WITH CROW BEADS

Sample length: 15.5 inches
Materials: (2) 7 foot lengths of 45 lb. test hemp
     (3) 9 mm crow beads (gray in the example)

This piece can also be wrapped twice around the wrist and worn as a bracelet.

Step 1:   Fold the two cords in half and tie a Slide Loop Clasp (shown as tied, with loop closed).
Step 2:   Tie a 4" long Left Square Knot Sinnet (about 23 Square Knots; a flat sinnet).
Step 3:   Tie a Switch Knot.
Step 4:   Tie a Left Square Knot.
Step 5:   String the first crow bead on the carriers.
Step 6:   Bring the knotters around either side of the bead and tie a Left Square Knot snugly beneath the bead.
Step 7:   Tie another Switch Knot.
Step 8:   Thread the carriers through the second crow bead.
Step 9:   Bring the knotters around either side of the bead and tie a Left Square Knot snugly beneath the bead.
Step 10:   Tie a third Switch Knot.
Step 11:   Thread the third crow bead on the carriers.
Step 12:   Bring the knotters around either side of the bead and tie a Left Square Knot snugly beneath the bead.
Step 13:   Add a Switch Knot
Step 14:   Tie a 4" long Left Square Knot Sinnet (about 23 Square Knots).
Step 15:   Add a Switch Knot.
Step 16:   Continue adding Square Knots until the desired length is reached (6 in the example).
Step 17:   Fold the carriers back (up) over the last square knot and tie 2 Square Knots over the folded carriers and the last knot.
Step 18:   Cut the carriers and the knotters and saturate the cut ends with glue.

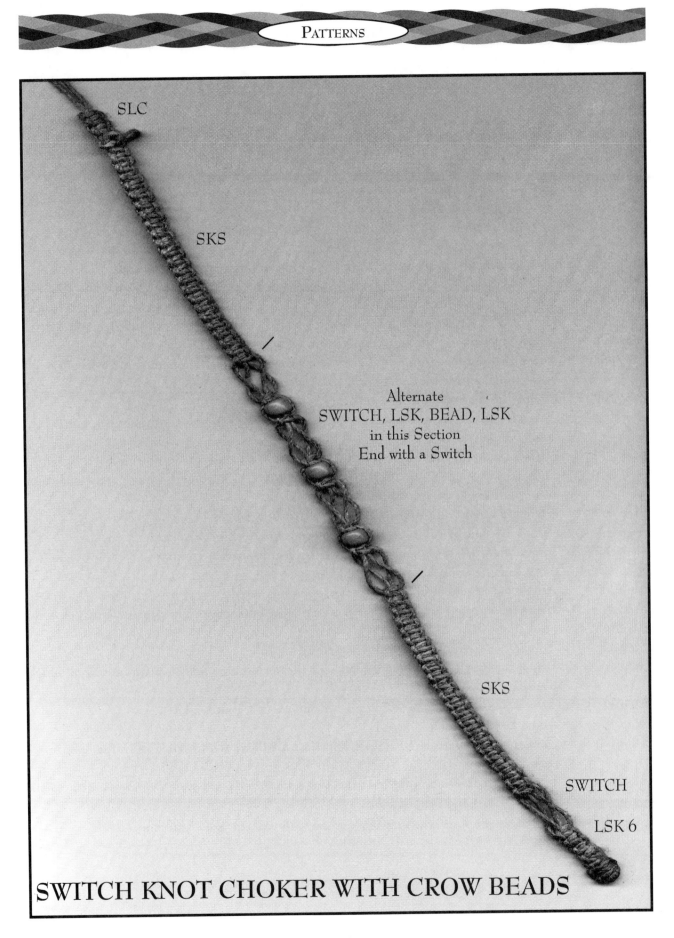

SLC

SKS

Alternate
SWITCH, LSK, BEAD, LSK
in this Section
End with a Switch

SKS

SWITCH

LSK 6

SWITCH KNOT CHOKER WITH CROW BEADS

45

SQUARE KNOT CHOKER WITH MUSHROOM FETISH

Sample length: 15 inches
Materials: (2) 7 foot lengths of black 45 lb. test hemp
    (2) 1/8 inch diam. spring tubes (nickel or silver in the example)
    (2) 5 or 6 mm beads (glossy black in the example)
    (1) jump ring
    (1) fetish or charm (8 mm x 12 mm mushroom in example)

This piece can also be wrapped twice around the wrist and worn as a bracelet.

Step 1:   Fold the two cords in half and tie a Slide Loop Clasp (shown as tied, with loop closed).

Step 2:   Tie a 4 3/4" long Square Knot Sinnet (about 27 Square Knots; a flat sinnet).

Step 3:   Thread the carriers through the first spring tube.

Step 4:   Bring the knotters around either side of the spring tube and tie a Square Knot snugly beneath the spring tube.

Step 5:   String the first bead on the carriers.

Step 6:   Bring the knotters around either side of the bead and tie a tight Square Knot below the bead.

Step 7:   Attach the jump ring to the fetish loop. Open and close the jump ring by twisting the ends sideways with a pair of needle-nosed pliers. Do not pull the ends away from each other as this will distort the shape of the ring.

Step 8:   String the jump ring to the top of the left knotter.

Step 9:   Tie a Left Square Knot, capturing the jump ring in the "hip" made by the left knotter. Leave a small loop in this knotter hip so the jump ring can move freely.

Step 10:   Tie a Square Knot.

Step 11:   String the second bead on the carriers.

Step 12:   Bring the knotters around either side of the bead and tie a tight Square Knot below the bead.

Step 13:   Thread the carriers through the second spring tube.

Step 14:   Bring the knotters around either side of the spring tube and tie a Square Knot snugly beneath the spring tube.

Step 15:   Tie a 4 3/4" long Square Knot Sinnet (about 27 Square Knots).

Step 16:   Add a Switch Knot.

Step 17:   Continue adding Square Knots until the desired length is reached (5 in the example).

Step 18:   Fold the carriers back (up) over the last square knot and tie a Square Knot over the folded carriers and the last knot.

Step 19:   Cut the carriers and the knotters and saturate the cut ends with glue.

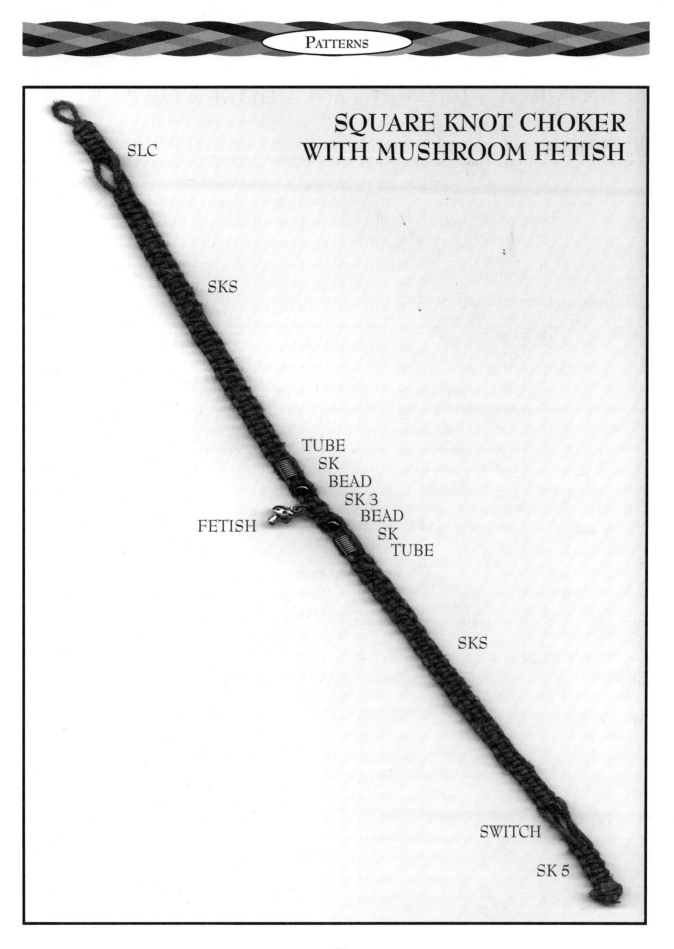

SQUARE KNOT CHOKER
WITH MUSHROOM FETISH

SLC

SKS

TUBE
SK
BEAD
SK 3
BEAD
SK
TUBE

FETISH

SKS

SWITCH

SK 5

CENTER BUTTON PRETZEL CHOKER OR ANKLET

Sample length: 13 inches
Materials: (2) 6 foot lengths of 45 lb. test hemp
                   (1) 5/8 inch, 4 hole button (shiny pink in the example)

Step 1:   Fold the two cords in half and tie a Slide Loop Clasp (shown with loop completely open, i.e. with switch knot collapsed).

Step 2:   Tie 5 Left Square Knots (a flat sinnet).

Step 3:   Tie 5 Left Half Knots (a spiral sinnet).

Step 4:   Tie 2 more Left Square Knots.

Step 5:   Tie 3 fairly tight Pretzel Knots.

Step 6:   Tie 2 Left Square Knots.

Step 7:   Tie 5 Right Half Knots.

Step 8:   Tie 2 more Left Square Knots.

Step 9:   String the carriers through the top two holes in the button, from back to front; bring the knotters around the front of the button and cross them under the carriers; string the carriers through the bottom two holes of the button, from front to back, and tighten over the knotters; bring the knotters around to the back of the button.

Step 10:   Tie 2 Left Square Knots.

Step 11:   Tie 5 Left Half Knots.

Step 12:   Tie 2 more Left Square Knots.

Step 13:   Tie 3 Pretzel Knots which are the same as those in Step 5.

Step 14:   Tie 2 Left Square Knots.

Step 15:   Tie 5 Right Half Knots.

Step 16:   Tie 6 Left Square Knots.

Step 17:   Add a Switch Knot.

Step 18:   Continue adding Left Square Knots until the desired length is reached.

Step 19:   Cut the carriers, put some glue on the cut ends, and tie one more Square Knot over the glued ends.

Step 20:   Tie 3 or 4 Square Knots back over the end of the pattern, glue the last knot, and cut off the knotters.

CENTER BUTTON PRETZEL
CHOKER OR ANKLET

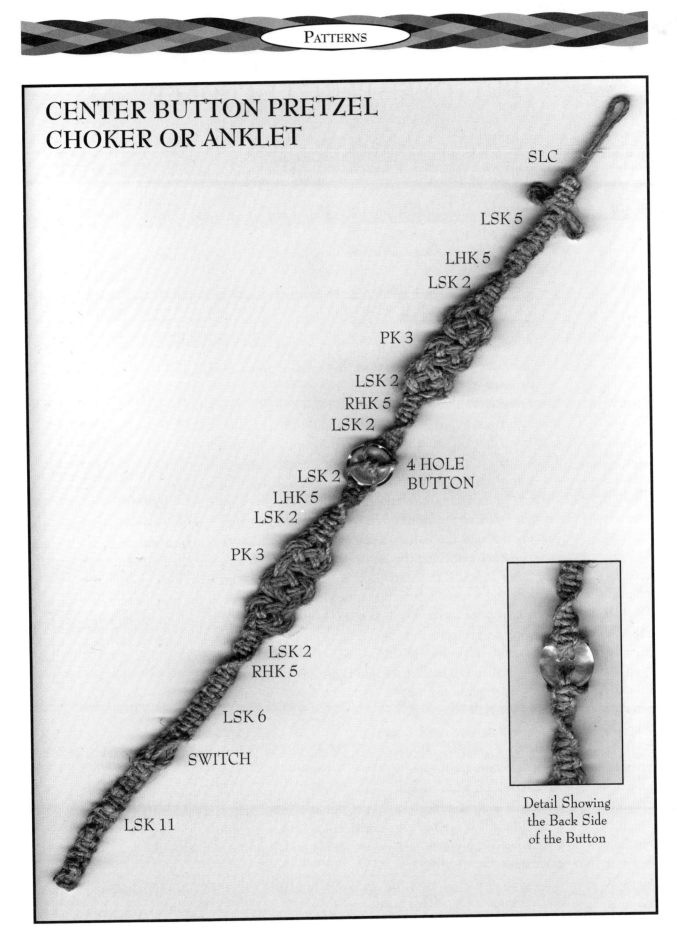

SLC

LSK 5

LHK 5
LSK 2

PK 3

LSK 2
RHK 5
LSK 2

4 HOLE
BUTTON

LSK 2
LHK 5
LSK 2

PK 3

LSK 2
RHK 5

LSK 6

SWITCH

LSK 11

Detail Showing
the Back Side
of the Button

BUTTONED PRETZEL CHOKER

Sample length: 17 inches
Materials: (2) 7.5 foot lengths of 45 lb. test hemp
    (1) 9/16 inch, 4 hole decorated metal button

Step 1: Fold the two cords in half and tie a Slide Loop Clasp (only bottom of Switch Knot in Slide Loop Clasp is shown).

Step 2: Tie 7 Left Square Knots (a flat sinnet).

Step 3: Tie a Switch Knot

Step 4: Tie 3 more Left Square Knots

Step 5: Tie 5 Right Half Knots (a spiral sinnet).

Step 6: Tie 2 more Left Square Knots.

Step 7: Tie 1 fairly tight Pretzel Knot.

Step 8: Leave a space of about 3/8 inch in the cords, with two cords (one knotter and one carrier) on each side of the project.

Step 9: Tie a second Pretzel Knot. All the Pretzel Knots in this project should be the same size and tightness.

Step 10: Leave another 3/8 inch space and tie a third Pretzel Knot.

Step 11: Leave another 3/8 inch space and tie a fourth Pretzel Knot.

Step 12: Tie 1 and 1/2 Square Knots (that is, a Square Knot followed by a Half Knot).

Step 13: String the carriers through the top two holes in the button, from back to front; bring the knotters around the front of the button and cross them under the carriers; string the carriers through the bottom two holes of the button, from front to back, and tighten over the knotters; bring the knotters down to the bottom of the button (see **Page 49**).

Step 14: Tie 1 and 1/2 Square Knots (that is, a Square Knot followed by a Half Knot).

Step 15: Tie a Pretzel Knot which is the same size and tightness as the previous Pretzel Knots.

Step 16: Leave a space of about 3/8 inch in the cords, with two cords on each side of the project.

Step 17: Tie a second Pretzel Knot.

Step 18: Leave another 3/8 inch space and tie a third Pretzel Knot.

Step 19: Leave another 3/8 inch space and tie a fourth Pretzel Knot.

Step 20: Tie 2 Left Square Knots.

Step 21: Tie 5 Left Half Knots.

Step 22: Tie 3 Left Square Knots.

Step 23: Add a Switch Knot.

Step 24: Tie 7 Left Square Knots.

Step 25: Add another Switch Knot.

Step 26: Continue adding sections of 7 Left Square Knots, followed by a Switch Knot, until the desired length is reached. End with 3 Left Square Knots.

Step 27: Cut the carriers, put some glue on the cut ends, and tie one more Square Knot over the glued ends.

Step 28: With each of the knotters, tie an Overhand Knot to the side of the piece. Add glue to each of these knots. This creates a "T hook" to catch in the Slide Loop Clasp.

BUTTONED PRETZEL CHOKER

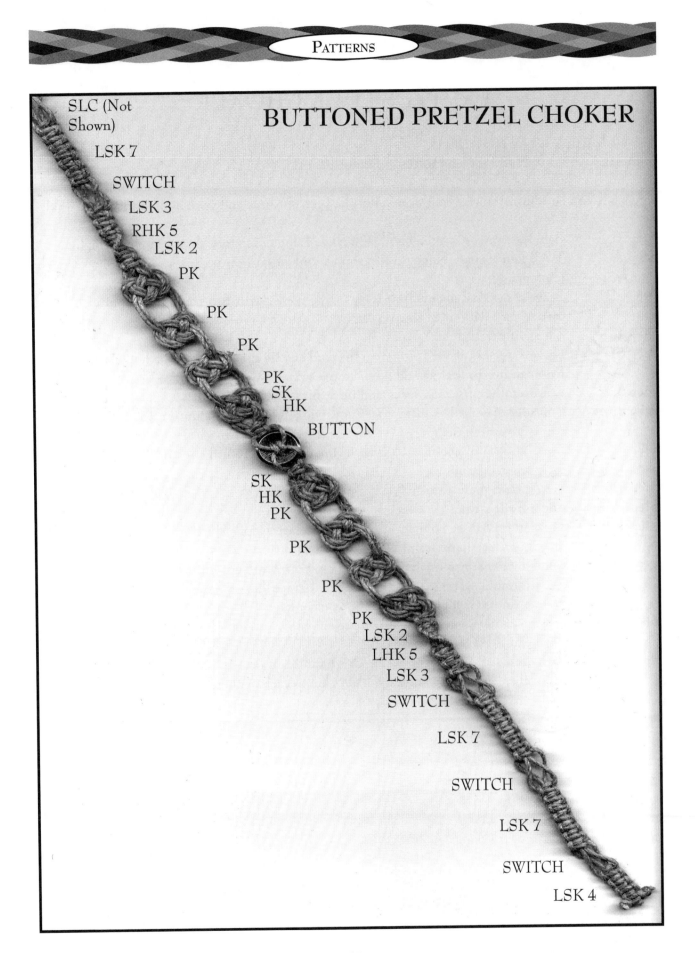

SLC (Not Shown)

LSK 7

SWITCH

LSK 3

RHK 5

LSK 2

PK

PK

PK

PK

SK

HK

BUTTON

SK

HK

PK

PK

PK

PK

LSK 2

LHK 5

LSK 3

SWITCH

LSK 7

SWITCH

LSK 7

SWITCH

LSK 4

PRETZEL LUCK CHOKER

Sample length: 17.75 inches
Materials: (2) 8 foot lengths of 45 lb. test hemp

This piece can be adjusted in size before and after the Pretzel Knots and worn as an armband or anklet.

Step 1:   Fold the two cords in half and tie a Slide Loop Clasp (not shown in photograph).

Step 2:   Tie 7 Left Square Knots (a flat sinnet; only a portion of this section is shown in the photograph).

Step 3:   Leave a space of about 3/16 inch in the cords, with two cords (one knotter and one carrier) on each side of the project.

Step 4:   Tie a fairly tight Pretzel Knot.

Step 5:   Leave a space of about 1/4 inch in the cord and tie a second Pretzel Knot. All the Pretzel Knots in this project should be the same size and tightness.

Step 6:   Leave another 1/4 inch space and tie a third Pretzel Knot.

Step 7:   Continue leaving 1/4 inch spaces and tying Pretzel Knots until a total of 13 Pretzel Knots have been tied.

Step 8:   Leave a space of about 3/16 inch in the cords and return the knotters and carriers to their original positions

Step 9:   Tie 7 Left Square Knots.

Step 10:   Add a Switch Knot.

Step 11:   The pattern will be completed once the Square Knots which mirror the "clasp" of the slide loop clasp have been tied. To add length to the project, alternate tying 7 Square Knots followed by a Switch Knot. The last Switch Knot must be followed by at least 3 Square Knots. The example in the photograph has two such sections, ending with 7 Square Knots (including the ending knot in Step 12).

Step 12:   Cut the carriers, put some glue on the cut ends, and tie one more Square Knot over the glued ends.

Step 13:   With each of the knotters, tie an Overhand Knot on either side of the last Square Knot, then cut off the knotters. Add glue to each of these knots. This creates a "T hook" to catch in the Slide Loop Clasp.

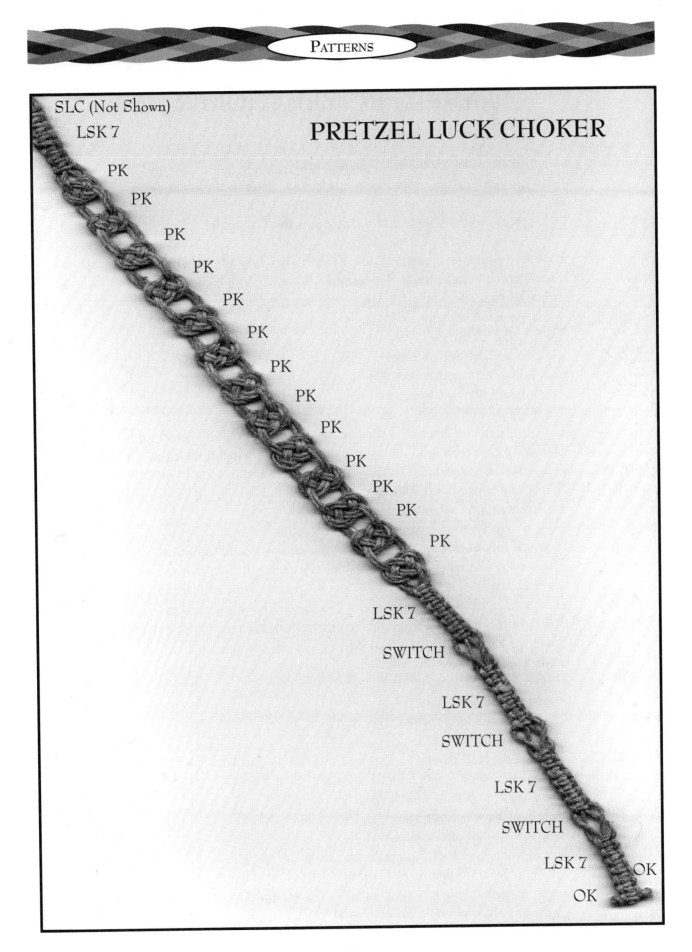

SLC (Not Shown)

LSK 7

PRETZEL LUCK CHOKER

PK

PK

PK

PK

PK

PK

PK

PK

PK

PK

PK

PK

PK

PK

LSK 7

SWITCH

LSK 7

SWITCH

LSK 7

SWITCH

LSK 7

OK

OK

FRAMED PRETZEL CHOKER

Sample length: 17 inches
Materials: (2) 7.5 foot lengths of 45 lb. test hemp
    (2) 9 mm, glass crow/pony beads (yellow luster in the example)

This piece can be shortened on both ends and worn as arm band or anklet.

Step 1:   Fold the two cords in half and tie a Slide Loop Clasp (shown as tied, with loop closed).
Step 2:   Tie 7 Left Square Knots (a flat sinnet).
Step 3:   Tie a short (1/2 inch) Switch Knot.
Step 4:   Tie 2 more Left Square Knots.
Step 5:   Tie 5 Left Half Knots (a spiral sinnet).
Step 6:   Tie 2 more Left Square Knots.
Step 7:   Tie another 1/2 inch Switch Knot.
Step 8:   Tie 2 more Left Square Knots.
Step 9:   Tie 5 Right Half Knots.
Step 10:   Tie 2 1/2 Left Square Knots (that is, two Square Knots followed by a Half Knot).
Step 11:   String the first crow bead on the two bead carriers, snug against the last knot tied.
Step 12:   Bring the knotters around either side of the bead and tie 2 1/2 Left Square Knots snug against the bottom of the bead.
Step 13:   Tie 1 fairly tight Pretzel Knot.
Step 14:   Tie a second Pretzel Knot, snug against the first, with no space between them. All the Pretzel Knots in this project should be the same size and tightness.
Step 15:   Tie as many Pretzel Knots as desired for the center of the pattern (there are a total of 5 in the example). The second half of the choker should be a mirror image of the first half.
Step 16:   Tie 2 1/2 Left Square Knots (again, two Square Knot followed by a Half Knot).
Step 17:   String the bead carriers through the second crow bead, snug against the last knot tied.
Step 18:   Bring the knotters around either side of the bead and tie 2 1/2 Left Square Knots snug against the bottom of the bead
Step 19:   Tie 5 Left Half Knots.
Step 20:   Tie 2 Left Square Knots.
Step 21:   Add a half inch long Switch Knot.
Step 22:   Tie 2 Left Square Knots.
Step 23:   Add 5 Right Half Knots.
Step 24:   Tie 2 Left Square Knots followed by a Left Half Knot.
Step 25:   Add another half inch Switch Knot.
Step 26:   Tie 7 Left Square Knots.
Step 27:   Add another half inch Switch Knot.
Step 28:   The addition of 3 Left Square Knots completes the pattern. However, if the length is not right, continue adding sections of 7 Left Square Knots, followed by a Switch Knot, until the desired length is reached (as in the example).
Step 29:   Cut the carriers, glue the cut ends, and tie one more Square Knot over the glued ends.

Step 30:   With each of the knotters, tie an Overhand Knot to the side of the piece. Add glue to each of these knots. This creates a "T hook" to catch in the Slide Loop Clasp.

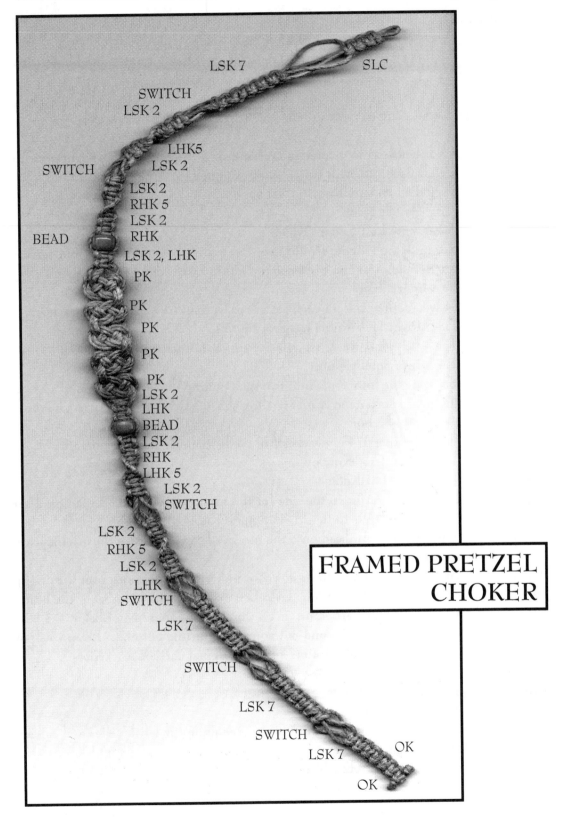

LSK 7   SLC

SWITCH
LSK 2

LHK5
SWITCH   LSK 2

LSK 2
RHK 5
LSK 2
BEAD   RHK
LSK 2, LHK
PK

PK

PK

PK

PK
LSK 2
LHK
BEAD
LSK 2
RHK
LHK 5
LSK 2
SWITCH

LSK 2
RHK 5
LSK 2
LHK
SWITCH

LSK 7

SWITCH

LSK 7

SWITCH   OK
LSK 7

OK

FRAMED PRETZEL CHOKER

PRETZEL CHOKER WITH CENTER BEAD

Sample length: 16.5 inches
Materials: (2) 8 foot lengths of 45 lb. test hemp
    (1) 3 inch length of 20 or 45 lb. test hemp (match bead hole size)
    (1) 6 mm bead (pearl gray in the example)

This piece can also be wrapped twice around the wrist and worn as a bracelet.

Step 1:   Fold the two 8 foot cords in half and tie a Slide Loop Clasp (shown with loop completely open, i.e. with switch knot collapsed).

Step 2:   Tie 7 Left Square Knots (a flat sinnet).

Step 3:   Tie a Switch Knot.

Step 4:   Tie 4 more Left Square Knots.

Step 5:   Separate the four cords into two sets, pulling the left knotter and carrier to the left and the right knotter and carrier to the right. Leave about 1/4" in each set of cords and then tie a Pretzel Knot.

Step 6:   Leave another 1/4" in each set of cords, then tie a Left Square Knot.

Step 7:   Tie 5 Right Half Knots (a spiral sinnet).

Step 8:   Tie 5 Left Half Knots (spirals the sinnet in the opposite direction).

Step 9:   Tie 1 Left Square Knot.

Step 10:   Separate the four cords into two sets, pulling the left knotter and carrier to the left and the right knotter and carrier to the right. Leave about 1/4" in each set of cords and then tie a Pretzel Knot.

Step 11:   Leave another 1/4" in each set of cords, then tie a Left Square Knot.

Step 12:   Tie 5 Right Half Knots.

Step 13:   Tie 5 Left Half Knots.

Step 14:   Leave the knotters to the side for the moment. Add the 3 inch bead carrier, using the original 2 carriers to tie 2 Left Square Knots to secure it.

Step 15:   Thread the center bead onto the new bead carrier. Leave the original carriers to the side for the moment.

Step 16:   Bring the original knotters down around the carrier knots, over the original carriers, and on either side of the bead. Tie a Left Square Knot snug below the bead.

Step 17:   Use the original carriers to tie a second Left Square Knot below the bead.

Step 18:   Cut the added carrier and put some glue on the cut end. From this point on, use the original knotters as knotters and the original carriers as carriers.

Step 19:   Tie 5 Right Half Knots.

Step 20:   Tie 5 Left Half Knots.

Step 21:   Tie 1 Left Square Knot.

Step 22:   Separate the four cords into two sets, pulling the left knotter and carrier to the left and the right knotter and carrier to the right. Leave about 1/4" in each set of cords and then tie a Pretzel Knot.

Step 23:   Leave another 1/4" in each set of cords, then tie a Left Square Knot.

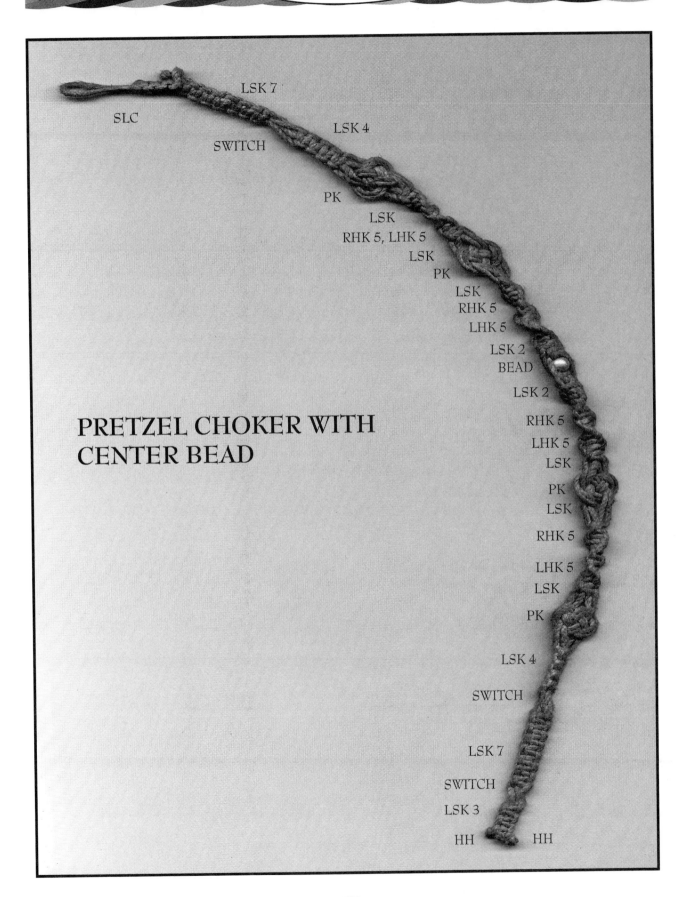

LSK 7

SLC

LSK 4

SWITCH

PK

LSK

RHK 5, LHK 5

LSK

PK

LSK

RHK 5

LHK 5

LSK 2

BEAD

LSK 2

RHK 5

LHK 5

LSK

PK

LSK

RHK 5

LHK 5

LSK

PK

LSK 4

SWITCH

LSK 7

SWITCH

LSK 3

HH   HH

PRETZEL CHOKER WITH CENTER BEAD

PRETZEL CHOKER WITH CENTER BEAD (Continued)

Step 24:   Tie 5 Right Half Knots.
Step 25:   Tie 5 Left Half Knots.
Step 26:   Tie 1 Left Square Knot.
Step 27:   Separate the four cords into two sets, pulling the left knotter and carrier to the left and the right knotter and carrier to the right. Leave about 1/4" in each set of cords and then tie a Pretzel Knot.
Step 28:   Leave another 1/4" in each set of cords, then tie a Left Square Knot.
Step 29:   Tie 3 more Left Square Knots.
Step 30:   Add a Switch Knot.
Step 31:   Tie 7 Left Square Knots, 1 Switch Knot and 3 Left Square Knots.
Step 32:   Cut the carriers and glue the ends (saturate the hemp fibers). Take the knotters individually and tie a Half Hitch to either side of the end of the piece. Glue these knots as well.

PHISH BONE CHOKER WITH FLAT CENTER BEAD

Sample Length: 16 inches
Materials:
       (2) 7 foot lengths of 45 lb. test hemp
       (1) 3.5 foot length of 45 lb. test hemp
       (1) 1 foot length of 20 lb. test hemp
       (1) 3/4 inch, 1 hole flat bead (red in the example)
       (12) small (3-4 mm) glass beads (6 red and 6 black in the example)

This piece can also be wrapped twice around the wrist and worn as a bracelet.

Step 1:   Fold the two 7 foot cords in half and tie a Slide Loop Clasp (shown with loop completely open, i.e. with switch knot collapsed).

Step 2:   Tie 9 Left Square Knots (a flat sinnet).

Step 3:   Add the 1 foot length of 20 lb. hemp as a bead carrier (between the original carriers), and, at the same time, add the 3.5 foot length of 45 lb. hemp as a new set of knotters. Use the original knotters to tie the Left Square Knot which secures the two new cords. There are now two sets of knotters (2 on the right and 2 on the left).

Step 4:   Pull the original carriers out to the sides as a third set of knotters, and use them to tie a Left Square Knot around the single thin bead carrier.

Step 5:   Begin tying the Phish Bone pattern. Bring the top set of knotters over (in front of) the two lower sets of knotters. Allow enough slack in the knotters to create the characteristic loop of the Phish Bone and tie a tight Square Knot around the single thin carrier (just below the last knot tied).

Step 6:   Bring the second set of knotters over (in front of) the third and first sets of knotters. Tie a tight Square Knot around the carrier, making sure the loops of the Phish Bone are the same size as the ones just tied.

Step 7:   Bring the third set of knotters over (in front of) the first and second sets of knotters. Check the size of the loops and tie a tight Square Knot around the carrier.

Step 8:   String one of the small beads (black) on the carrier. Repeat Step 5, tying the Square Knot just below the bead. Make the loops slightly larger to accommodate and showcase the beads. Maintain this larger loop size until after the last bead has been added.

Step 9:   String another small bead (red) on the carrier. Repeat Step 6, tying the Square Knot just below the new bead

Step 10:   String a third small bead (black) on the carrier. Repeat Step 7, tying the Square Knot just below the third bead.

Step 11:   Repeat Step 5.

Step 12:   String a small bead (red) on the carrier. Repeat Step 6, tying the Square Knot just below the new bead.

Step 13:   String another small bead (black) on the carrier. Repeat Step 7, tying the Square Knot just below the bead.

Step 14:   String a sixth small bead (red) on the carrier. Repeat Step 5, tying the Square Knot just below the bead.

Step 15:   Repeat Step 6.

Step 16:   Repeat Step 7.

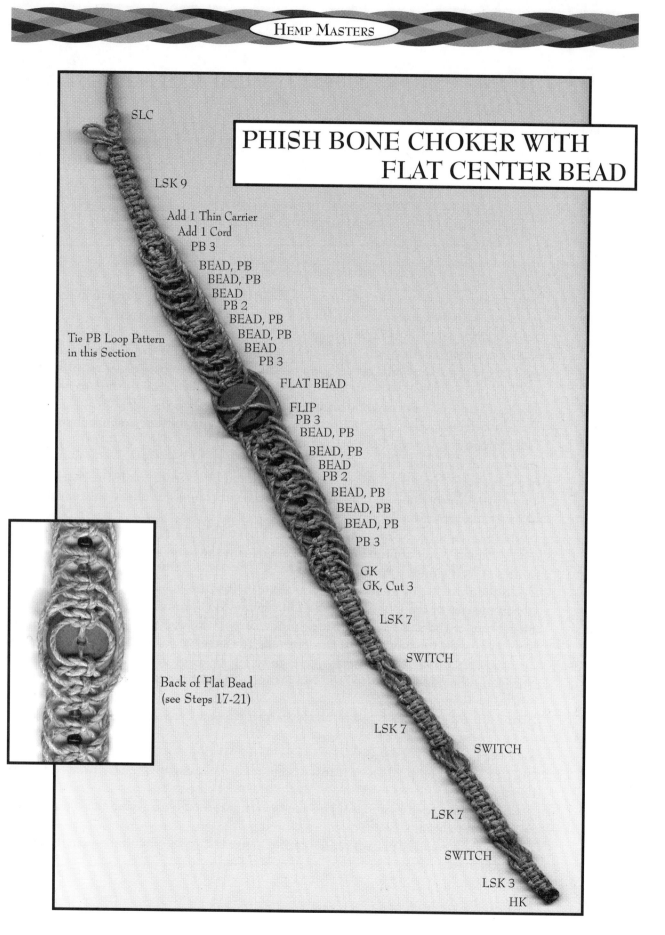

PHISH BONE CHOKER WITH FLAT CENTER BEAD

SLC

LSK 9

Add 1 Thin Carrier
Add 1 Cord
PB 3

BEAD, PB
BEAD, PB
BEAD
PB 2
BEAD, PB
BEAD, PB
BEAD
PB 3

Tie PB Loop Pattern
in this Section

FLAT BEAD

FLIP
PB 3
BEAD, PB

BEAD, PB
BEAD
PB 2
BEAD, PB
BEAD, PB
BEAD, PB
PB 3

GK
GK, Cut 3

LSK 7

SWITCH

Back of Flat Bead
(see Steps 17-21)

LSK 7

SWITCH

LSK 7

SWITCH

LSK 3
HK

Step 17:   Thread a small loop of the bead carrier up through the hole in the large flat bead (from back to front). Using what is now the middle set of knotters, cross the knotters in front of the flat bead, threading each one through the small carrier loop in the middle. Pull the knotters snug around the top of the bead, then tighten the carrier loop to hold them in place. Leave these knotters untied for the moment.

Step 18:   Flip the pattern over (or face down), so that the back of the flat bead faces up. The second half of the pattern will be tied with the choker in this position.

Step 19:   Take the bottom set of knotters, make the Phish Bone loops just small enough so they will not show on the front side, and tie a Square Knot on the carrier, just below the hole in the bead.

Step 20:   Bring the crossed middle knotters around the bottom of the flat bead, over (in front of) the knotters just used, and tie a Square Knot.

Step 21:   Bring the top set of knotters down, in front of (over) all of the cords and knots around the flat bead. Make Phish Bone loops that are large enough to frame the flat bead and can be seen from either side of the pattern. Tie a Square Knot just below the last knot tied.

Step 22:   String a small bead (red) on the carrier. Using what is now the top set of knotters, repeat Step 5, tying the Square Knot just below the small bead.

Step 23:   String another small bead (black) on the carrier. Using the next set of knotters, repeat Step 6, tying the Square Knot below the bead just added.

Step 24:   String a third small bead (red) on the carrier. Using the third set of knotters, repeat Step 7, tying the Square Knot below the new bead.

Step 25:   Repeat Step 5.

Step 26:   String a small bead (black) on the carrier and repeat Step 6, tying the Square Knot below the black bead.

Step 27:   String a small bead (red) on the carrier and repeat Step 7, tying the Square Knot below the bead.

Step 28:   String a small bead (black) on the carrier and repeat step 5, tying the Square Knot below this last bead.

Step 29:   Repeat Steps 6 and 7.

Step 30:   Repeat Step 5.

Step 31:   Bring the next set of knotters down into position, but instead of tying a knot, place them parallel to the bead carrier. Add a bit of glue and use the last set of knotters from Step 29 to tie a Gathering Knot around these three cords.

Step 32:   Repeat Step 31, bringing the next set of knotters in as carriers. Add some glue and use the same set of knotters as in Step 31 to tie a Gathering Knot around the five cords now in the center.

Step 33:   Cut the thin bead carrier and the two carriers pulled into the center in Step 31.

Step 34:   Tie 7 Left Square Knots, using the knotters from Steps 31-32.

Step 35:   Tie a Switch Knot.

Step 36:   Flip the pattern over, so that the center bead is face up. The pattern will be completed once the Square Knots which mirror the "clasp" of the slide loop clasp have been tied. To add length to the project, alternate tying 7 Square Knots followed by a Switch Knot. The last Switch Knot must be followed by at least 3 Square Knots.

Step 37:   End by cutting the two middle carriers and putting some glue on the cut ends. Tie a Half Knot over the glued ends, then cut off the knotters.

SMALL PHISH BONE CHOKER

Sample Length: 17 inches
Materials: (2) 7 foot lengths of 45 lb. test hemp
                  (1) 6 foot length of 45 lb. test hemp
                  (1) 1 foot length of 20 lb. test hemp
                  (1) crystal center bead (clear oval in the example)
                  (12) small (3-4 mm) glass beads (6 light blue and 6 dark blue in the example)

This piece can be adjusted in size before and after the Phish Bone and worn as an armband.

Step 1:   Fold the two 7 foot cords in half and tie a Slide Loop Clasp (not shown in photograph).
Step 2:   Tie 7 Left Square Knots (a flat sinnet).
Step 3:   Tie a Switch Knot.
Step 4:   Add the 1 foot length of 20 lb. hemp as a third bead carrier (between the original carriers). Use 2 Left Square Knots to secure it in place.
Step 5:   Tie 5 Right Half Knots (a spiral sinnet).
Step 6:   Add the 6 foot length of 45 lb. hemp across the carriers as a new set of knotters. Leave the ends of the new cord out to the sides. Use the original knotters to tie the 2 Left Square Knots which secure the new cord. There are now two sets of knotters (2 on the right and 2 on the left).
Step 7:   Pull the original carriers out to the sides as a third set of knotters, and use them to tie a Left Square Knot around the single thin bead carrier.
Step 8:   String the beads on the single thin (20 lb.) bead carrier in the following order: 6 small beads, alternating light and dark; 1 center bead; 6 small beads, alternating dark and light (a mirror image of the first six). Tie a loose Overhand Knot in the end of the bead carrier so the beads can not fall off. Allow the beads to slide down to the overhand knot. Do not tie the beads into the pattern until indicated.
Step 9:   Begin tying the Phish Bone pattern. To do this, bring the top set of knotters over (in front of) the two lower sets of knotters. Allow enough slack in the knotters to create the characteristic loop of the Phish Bone, but keep the loops small for this pattern. Tie a tight Square Knot around the single thin carrier (just below the last knot tied). Pull on the loops at the same time as the knotters to get this knot extra tight.
Step 10: Bring the second set of knotters over (in front of) the third and first sets of knotters. Tie a tight Square Knot around the carrier, making sure the loops of the Phish Bone are the same size as the ones just tied. Again, pull the loops and the knotters at the same time to tighten this knot.
Step 11: Bring the third set of knotters over (in front of) the first and second sets of knotters. Check the size of the loops and tie a tight Square Knot around the carrier. Repetition of these three steps (9, 10 and 11) creates the Phish Bone effect; variations in loop size and bead placement make each piece unique.
Step 12: Bring one of the small beads (light) up the carrier to just below the last knot tied. Con-

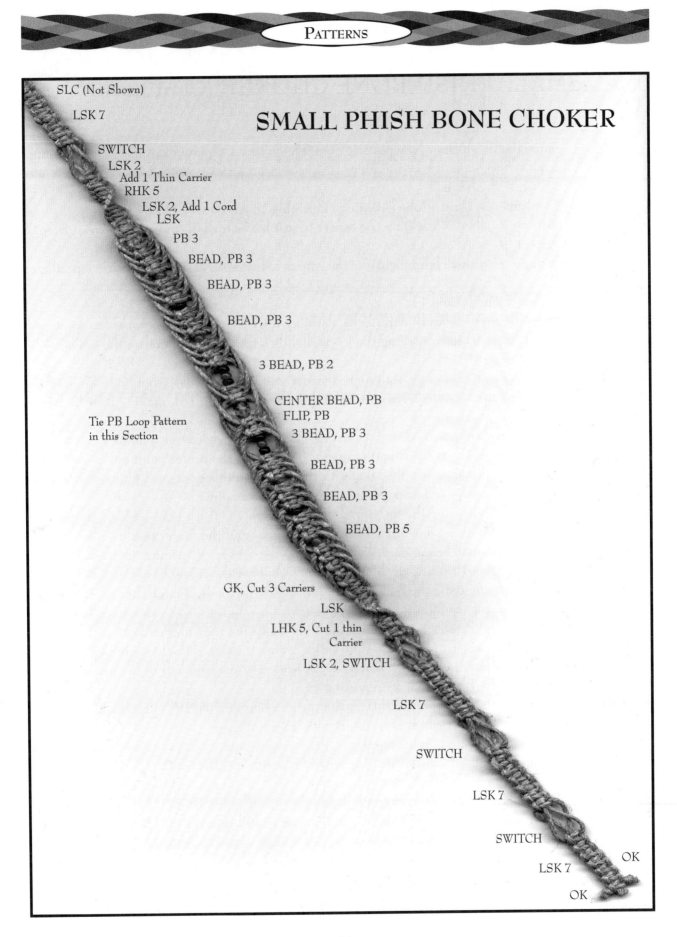

SMALL PHISH BONE CHOKER

SLC (Not Shown)

LSK 7

SWITCH
LSK 2
Add 1 Thin Carrier
RHK 5
LSK 2, Add 1 Cord
LSK
PB 3
BEAD, PB 3

BEAD, PB 3

BEAD, PB 3

3 BEAD, PB 2

CENTER BEAD, PB
FLIP, PB
3 BEAD, PB 3

Tie PB Loop Pattern
in this Section

BEAD, PB 3

BEAD, PB 3

BEAD, PB 5

GK, Cut 3 Carriers

LSK

LHK 5, Cut 1 thin
Carrier

LSK 2, SWITCH

LSK 7

SWITCH

LSK 7

SWITCH

OK

LSK 7

OK

63

SMALL PHISH BONE CHOKER (Continued)

tinue the Phish Bone by repeating Step 9 and tying a Square Knot just below and snug against the bead. Make the loops in this section slightly larger to accommodate and showcase the beads.

Step 13: Continue the Phish Bone pattern by repeating Steps 10 and 11.

Step 14: Bring a small bead (dark) up the carrier to just below the last knot tied. Repeat Step 9.

Step 15: Repeat Steps 10 and 11.

Step 16: Bring a third small bead (light) up the carrier to below the last knot tied and repeat Step 9.

Step 17: Repeat Steps 10 and 11.

Step 18: Bring three small beads (dark, light, dark) up the carrier to just below the last knot tied. Repeat Step 9, again making the loops slightly larger to accommodate the beads.

Step 19: Repeat Step 10.

Step 20: Bring the center bead (crystal) up the carrier, snug below the last knot tied. Repeat Step 11 (using the knotters from above the three beads in Step 18), making this the largest loop in the pattern so far.

Step 21: Flip the piece of jewelry over (face down). This step is needed to make the second half of the pattern match the first half (a mirror image from the center point out to each end). The rest of the pattern will be tied with the choker in this position. This technique will work for any piece of hemp jewelry that has a center point.

Step 22: Repeat Step 9 (using the knotters from just below the three beads, which are now the first set of knotters), making a fairly large loop around the center bead.

Step 23: Bring three small beads (dark, light, dark) up the carrier. Continue the Phish Bone by repeating Step 10 (using the knotters from just above the center bead which are now the second set) and tying the Square Knot just below the beads. Match the size of this loop to the one in Step 20.

Step 24: Continue the Phish Bone by repeating Step 11 (using the knotters from just below the center bead, which are now the third set). Match the size of the loops in this section to the size of those in the same (single bead) section of the first half of the pattern.

Step 25: Continue the Phish Bone by repeating Step 9.

Step 26: Bring another small bead (light) up the carrier. Repeat Step 10, tying the Square Knot below the bead just added.

Step 27: Repeat Step 11 and then Step 9.

Step 28: Bring a small bead (dark) up the carrier and repeat Step 10, tying the Square Knot below the dark bead just added.

Step 29: Repeat Step 11 and then Step 9.

Step 30: Bring the final small bead (light) up the carrier and repeat step 10, tying the Square Knot below this last bead.

Step 31: Repeat Step 11. The loops in this section should be the same size as those in the beginning section of the Phish Bone.

SMALL PHISH BONE CHOKER (Continued)

Step 32:   Repeat Steps 9 and 10.

Step 33:   Place the knotters from Step 32 parallel to the bead carrier, below the last knot tied. Add a bit of glue and use the knotters from Step 31 to tie a Square Knot around all three cords.

Step 34:   Place the knotters from step 33 parallel to the bead carriers, below the knot just tied. Add a bit of glue and use the knotters from Step 32 to tie a Gathering Knot, consisting of two Square Knots, around all five cords.

Step 35:   Cut the extra set of knotters (the 45 lb. cords pulled into the center as carriers in Step 33).

Step 36:   Use the remaining knotters (the ones used in Step 34) to tie a Left Square Knot around the two remaining carriers.

Step 37:   Tie 5 Left Half Knots.

Step 38:   Cut the thin carrier, put a dab of glue on the end and tie two Left Square Knots over the end to secure it from slipping.

Step 39:   Tie a Switch Knot.

Step 40:   Tie 7 Left Square Knots.

Step 41:   Tie another Switch Knot.

Step 42:   The pattern will be completed once the Square Knots which mirror the "clasp" of the slide loop clasp have been tied. To add length to the project, alternate tying 7 Square Knots followed by a Switch Knot. The last Switch Knot must be followed by at least 3 Square Knots. The example in the photo has one such repetition, ending with 7 Square Knots.

Step 43:   End by cutting the two middle carriers and putting some glue on the cut ends. With each of the knotters, tie an Overhand Knot on either side of the last Square Knot, then cut off the knotters. Add glue to each of these knots. This creates a "T hook" to catch in the Slide Loop Clasp.

BUTTONED PHISH BONE ARM OR ANKLE BAND

Sample Length: 17 inches
Materials: (2) 8 foot lengths of 45 lb. test hemp
                   (1) 7 foot length of 45 lb. test hemp
                   (1) 2 foot length of 20 lb. test hemp
                   (1) half inch square button with loop on the back (blue in the example)
                   (12) 3-4 mm glass beads (6 light blue and 6 black in the example)

This piece can also be worn as a choker or it can be wrapped twice around the wrist and worn as a bracelet.

Step 1:   Fold the two 8 foot cords in half and tie a Slide Loop Clasp (not shown).
Step 2:   Tie 7 Left Square Knots (a flat sinnet).
Step 3:   Tie a short (half inch) Switch Knot.
Step 4:   Add the 2 foot length of 20 lb. hemp as a bead carrier (between the original carriers) and tie two Left Square Knots to secure the new cord.
Step 5:   Tie 5 Right Half Knots (a spiral sinnet).
Step 6:   Tie 2 Left Square Knots.
Step 7:   Add the 7 foot length of 45 lb. hemp as a new set of knotters. Place the cord across the carriers and use the original knotters to tie a Left Square Knot to secure the new cord. There are now two sets of knotters (2 on the right and 2 on the left).
Step 8:   Pull the original carriers out to the sides as a third set of knotters, and use them to tie a Left Square Knot around the single thin bead carrier (snug to the last knot tied).
Step 9:   String six of the twelve small glass beads on the single thin bead carrier. If two colors are being used, string the first six in the following order: dark, light, dark, light, dark, light. String the button through the hole (loop) on the back. String the second six beads in the reverse order: light, dark, light, dark, light, dark. Tie a loose overhand knot in the bottom of the bead carrier so the beads won't fall off and allow the beads to slide to the bottom of the cord.
Step 10:   Begin tying the Phish Bone pattern. Start by bringing the top set of knotters over (in front of) the two lower sets of knotters. Allow some slack in the knotters to create the characteristic loop of the Phish Bone and tie a tight Square Knot around the single thin carrier (just below the last knot tied). This first Phish loop should be fairly small because in this pattern the size of the loops should gradually increase all the way to the center point. The loops around the button at the center should be the largest of all.
Step 11:   Bring the second set of knotters over (in front of) the third and first sets of knotters. Tie a tight Square Knot around the carrier, increasing the loop size of the Phish Bone just a bit (also be sure that the loops are the same on either side of the carrier).
Step 12:   Bring the third set of knotters over (in front of) the first and second sets of knotters. Check the size of the loops and tie a tight Square Knot around the carrier. These three steps are the ones that create the Phish Bone and they will be repeated throughout this section of the pattern.

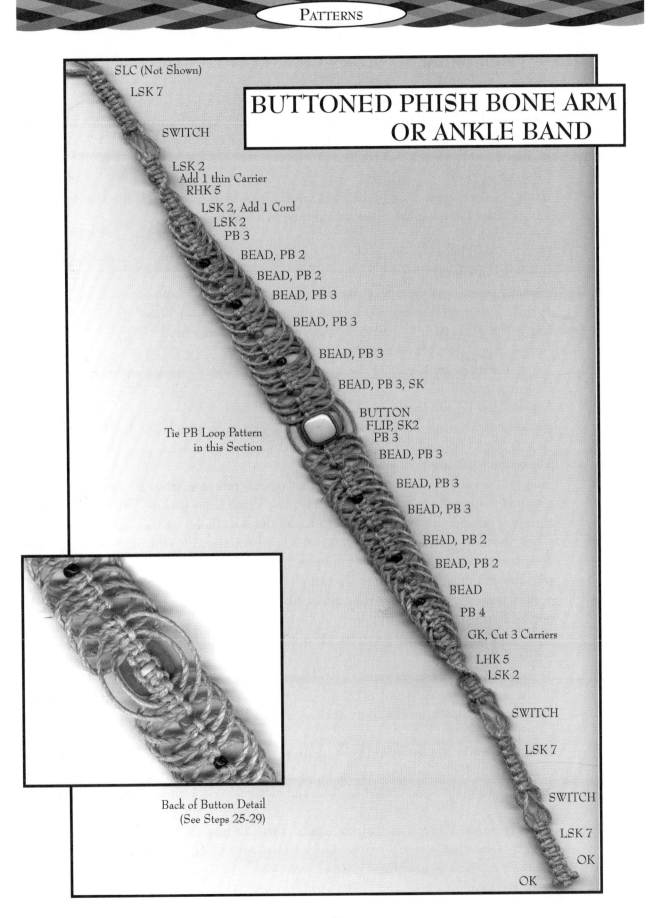

BUTTONED PHISH BONE ARM OR ANKLE BAND

SLC (Not Shown)

LSK 7

SWITCH

LSK 2
Add 1 thin Carrier
RHK 5
LSK 2, Add 1 Cord
LSK 2
PB 3
BEAD, PB 2
BEAD, PB 2
BEAD, PB 3
BEAD, PB 3
BEAD, PB 3
BEAD, PB 3, SK

BUTTON
FLIP, SK2
PB 3
BEAD, PB 3
BEAD, PB 3
BEAD, PB 3
BEAD, PB 2
BEAD, PB 2
BEAD
PB 4
GK, Cut 3 Carriers
LHK 5
LSK 2

SWITCH

LSK 7

SWITCH

LSK 7

OK

OK

Tie PB Loop Pattern
in this Section

Back of Button Detail
(See Steps 25-29)

BUTTONED PHISH BONE ARM OR ANKLET BAND (Continued)

Step 13:   Bring one of the small beads (black) up the carrier to just below the last knot tied. Continue the Phish Bone by repeating Step 10 and tying a Square Knot just below and snug against the bead. Make the loops slightly larger to accommodate and showcase the beads.

Step 14:   Repeat Step 11.

Step 15:   Bring another small bead (light blue) up the carrier to just below the last knot tied. Repeat Step 12, tying the Square Knot just below the new bead.

Step 16:   Repeat Step 10.

Step 17:   Bring a small bead (black) up the carrier. Repeat Step 11, tying the Square Knot just below the new bead.

Step 18:   Repeat Step 12 and then Step 10.

Step 19:   Bring another small bead (light blue) up the carrier to just below the last knot tied. Repeat Step 11, tying the Square Knot just below the new bead.

Step 20:   Repeat Step 12 and then Step 10.

Step 21:   Bring a small bead (black) up the carrier. Repeat Step 11, tying the Square Knot just below the new bead.

Step 22:   Repeat Step 12 and then Step 10.

Step 23:   Bring another small bead (light blue) up the carrier to just below the last knot tied. Repeat Step 11, tying the Square Knot just below the new bead.

Step 24:   Repeat Step 12 and then Step 10.

Step 25:   Now the center button is added. Using the last knotters from Step 24 tie a Square Knot around the single bead carrier (do not make any Phish Bone loops). This accommodates the space between the edge of the button and the loop (hole) on the back of the button.

Step 26:   Bring the button up the carrier against the last knot tied. Flip the pattern over (or face down), so that the back of the button faces up. The second half of the pattern will be tied with the arm or ankle band in this position. Using the same set of knotters, bring them around (outside) the edges of the button hole and tie a Square Knot snug to secure the button in place.

Step 27:   Still using the same set of knotters, tie one more Square Knot around the bead carrier (again without any Phish Bone loops).

Step 28:   To begin framing the center button (still face down), bring the knotters closest to the top of the button over (in front of) the knotters just used. Make Phish Bone loops big enough to be seen from the front of the pattern. Tie a Square Knot just below the last knot tied.

Step 29:   Bring the top set of knotters down, in front of (over) the other two sets of knotters. Make Phish Bone loops that are large enough to frame the button and the loops from Step 32 (make these the largest loops in the pattern). Tie a Square Knot below the last knot tied.

Step 30:   Begin tying the Phish Bone pattern again. First, bring a small bead (light blue) up the carrier. Take what is now the top set of knotters over (in front of) the two lower sets of knotters and tie a Square Knot just below the bead. Remember to match the size of the

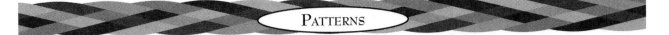

BUTTONED PHISH BONE ARM OR ANKLET BAND (Continued)

Phish Bone loops to those on the first half of the pattern.

Step 31:   Bring the second set of knotters over (in front of) the third and first sets of knotters. Tie a tight Square Knot around the carrier, decreasing the loop size of the Phish Bone just a bit.

Step 32:   Bring the third set of knotters over (in front of) the first and second sets of knotters. Check the size of the loops and tie a tight Square Knot around the carrier. These three will be repeated throughout this section of the pattern.

Step 33:   Bring another small bead (black) up the carrier. Using the top set of knotters, repeat Step 30, tying the Square Knot below the bead just added.

Step 34:   Repeat Steps 31 and 32.

Step 35:   Bring a third small bead (light blue) up the carrier. Repeat Step 30, tying the Square Knot below the new bead.

Step 36:   Repeat Steps 31 and 32.

Step 37:   Bring a fourth small bead (black) up the carrier and repeat Step 34, tying the Square Knot below the new bead.

Step 38:   Repeat Step 31.

Step 39:   Bring a fifth small bead (light blue) up the carrier and repeat Step 32, tying the Square Knot below the bead.

Step 40:   Repeat Step 30, without a bead.

Step 41:   Bring the last small bead (black) up the carrier and repeat Step 31, tying the Square Knot below the bead.

Step 42:   Repeat Step 32, then Step 33, without a bead and finally Step 31.

Step 43:   Bring the last set of knotters used down into the carrier. Using the top set of knotters, tie a Square Knot around the three carriers, then place these knotters parallel to the other carriers. Add a bit of glue and use the last set of knotters to tie a Gathering Knot around these five cords.

Step 44:   Cut the thin bead carrier and the extra set of knotters pulled into the center in Step 43. This leaves two knotters and two carriers, all 45 lb. test. Continue tying the pattern over these cut and glued ends.

Step 45:   Tie 5 Left Half Knots, using the knotters from Step 43.

Step 46:   Tie 2 Left Square Knots.

Step 47:   Tie a short (half inch) Switch Knot.

Step 48:   Tie 7 Left Square Knots.

Step 49:   Add a Switch Knot.

Step 50:   The pattern will be completed once the Square Knots which mirror the "slide knots" of the slide loop clasp have been tied. To add length to the project, alternate tying 7 Square Knots followed by a Switch Knot. The last Switch Knot must be followed by at least 3 Square Knots.

Step 51:   Cut the carriers, put some glue on the cut ends, and tie one more Square Knot over the glued ends.

Step 52:   With each of the knotters, tie an Overhand Knot to the side of the piece. Add glue to each of these knots. This creates a "T hook" to catch in the Slide Loop Clasp.

LOOPED PHISH BONE CHOKER OR ANKLET

Sample Length: 15 inches
Materials: (2) 7 foot lengths of 45 lb. test hemp
(1) 6 foot length of 45 lb. test hemp
(1) 2 foot length of 20 lb. test hemp
(7) beads of graduated sizes (10 mm center, 8 mm on either side of center, and 5 mm on both ends, marbled tan in the example)

This piece can also be wrapped twice around the wrist and worn as a bracelet.

Step 1: Fold the two 7 foot cords in half and tie a Slide Loop Clasp (shown as tied, with loop closed).
Step 2: Tie 9 Square Knots (a flat sinnet).
Step 3: Add the 2 foot length of 20 lb. hemp as a bead carrier (between the original carriers) and tie four more Square Knots to secure the new cord and complete the sinnet.
Step 4: Add the 6 foot length of 45 lb. hemp as a new set of knotters. Place the cord across the carriers and use the original knotters to tie a Square Knot to secure the new cord. Leave the ends of the new cord out to the sides. There are now two sets of knotters (2 on the right and 2 on the left).
Step 5: Pull the original carriers out to the sides as a third set of knotters, and use them to tie a Square Knot around the single thin bead carrier (snug against the last knot tied).
Step 6: String the beads on the single thin bead carrier. Place the largest bead in the center, with the smaller beads equally distributed on either side of it (5 mm, 5 mm, 8 mm, 10 mm, 8 mm, 5 mm, 5 mm in the example). Tie a loose overhand knot in the bottom of the bead carrier so the beads won't fall off and allow the beads to slide to the bottom of the cord.
Step 7: Begin tying the Phish Bone pattern. Start by bringing the top set of knotters over (in front of) the two lower sets of knotters. Allow some slack in the knotters to create the characteristic loop of the Phish Bone and tie a tight Square Knot around the single thin carrier (just below the last knot tied). This first Phish loop should be fairly small because in this pattern the size of the loops should gradually increase until the looped center section is reached.
Step 8: Bring the second set of knotters over (in front of) the third and first sets of knotters. Tie a tight Square Knot around the carrier, increasing the loop size of the Phish Bone just a bit (also be sure that the loops are the same on either side of the carrier).
Step 9: Bring the third set of knotters over (in front of) the first and second sets of knotters. Check the size of the loops and tie a tight Square Knot around the carrier. These three steps are the ones that create the Phish Bone and they will be repeated to make this section of the pattern.
Step 10: Repeat Steps 7, 8 and 9.
Step 11: Repeat Step 7.
Step 12: Bring one of the small beads (5 mm) up the carrier to just below the last knot tied. To

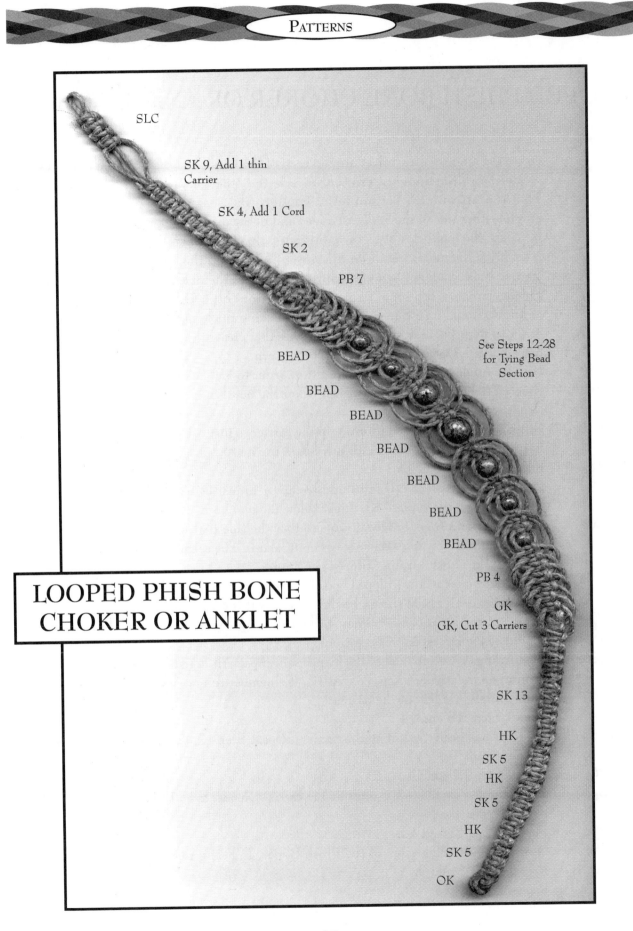

SLC

SK 9, Add 1 thin
Carrier

SK 4, Add 1 Cord

SK 2

PB 7

See Steps 12-28
for Tying Bead
Section

BEAD

BEAD

BEAD

BEAD

BEAD

BEAD

BEAD

PB 4

GK

GK, Cut 3 Carriers

SK 13

HK
SK 5
HK
SK 5
HK
SK 5
OK

LOOPED PHISH BONE
CHOKER OR ANKLET

LOOPED PHISH BONE CHOKER OR ANKLET (Cont'd)

create the looped saucer effect around each of the beads, use the same technique used to create the Phish Bone pattern, but reverse the order in which the knotters are used. That is, start with the bottom set of knotters (the ones used in Step 11), bring them around either side of the bead just added and tie a Square Knot under the bead to secure it. Make the loops closest to the beads fairly small.

Step 13:   Next bring the middle set of knotters over (in front of) the knotters just used, and tie a Square Knot, making the loops a little bigger than the ones in Step 12.

Step 14:   Finally, bring the top set of knotters over (in front of) both the bottom and middle sets of knotters, and tie a Square Knot, again making the loops bigger than the ones in the previous step. These three steps are the ones that create the saucer effect and they will be repeated to complete this section of the pattern.

Step 15:   Bring another small bead (5 mm) up the carrier to just below the last knot tied. Repeat Step 12, tying the Square Knot just below the new bead.

Step 16:   Repeat Steps 13 and 14.

Step 17:   Bring a medium size bead (8 mm) up the carrier to just below the last knot tied. Repeat Step 12, tying the Square Knot just below the new bead.

Step 18:   Repeat Steps 13 and 14.

Step 19:   Bring the center bead (10 mm) up the carrier to just below the last knot tied. Repeat Step 12, tying the Square Knot just below the new bead.

Step 20:   Flip the pattern over (or face down), so that the side that was up when all the previous knots were tied now faces the table or craft board. The second half of the pattern will be tied with the choker in this "flipped" position.

Step 21:   With the choker in the "flipped" position, repeat Step 13. Remember that the loops around the center bead should be the largest in this creation.

Step 22:   Repeat Step 14 (the choker remains "flipped"). Again, the loops around the center bead should be the largest in this creation.

Step 23:   Begin tying the second half of the looped saucer pattern. First, bring a medium size bead (8 mm) up the carrier. Repeat Step 12. Remember to match the size of these loops to those on the first half of the pattern.

Step 24:   Repeat Steps 13 and 14.

Step 25:   Bring a small bead (5 mm) up the carrier. Repeat Step 12, tying the Square Knot below the new bead.

Step 26:   Repeat Step 13 and 14.

Step 27:   Bring another small bead (5 mm) up the carrier. Repeat Step 12, tying the Square Knot below the new bead.

Step 28:   Repeat Steps 13 and 14.

Step 29:   Now tie the second section of the Phish Bone. To begin, repeat Step 7.

Step 30:   Repeat Steps 8 and 9.

Step 31:   Repeat Step 7.

LOOPED PHISH BONE CHOKER OR ANKLET (Cont'd)

Step 32:   Bring the set of knotters from Step 31 down into position, but instead of tying a knot, place them parallel to the bead carrier. Add a bit of glue and use the last set of knotters from Step 30 to tie a Gathering Knot around these three cords.

Step 33:   Repeat Step 32, bringing the first set of knotters from Step 30 in as carriers. Add some glue and use the second set of knotters from Step 30 to tie a Gathering Knot around the five cords now in the center.

Step 34:   Cut the thin bead carrier and the extra set of knotters pulled into the center in Step 32. This leaves two knotters and two carriers, all 45 lb. test. Continue tying the pattern over these cut and glued ends.

Step 35:   Tie 13 Square Knots, using the knotters from Step 31.

Step 36:   This example features a Ripple End for adjustable sizing. Tie a Half Knot without the carriers *on top* of the last Square Knot tied. This creates a bump for the loop of the slide loop clasp.

Step 37:   If desired, add a Switch Knot to match the beginning of the pattern (not shown in the example).

Step 38:   Tie 5 Square Knots.

Step 39:   Tie a Half Knot (without the carriers) on top of the last Square Knot tied.

Step 40:   Tie 5 Square Knots.

Step 41:   Tie a Half Knot (without the carriers) on top of the last Square Knot tied.

Step 42:   Tie 5 Square Knots.

Step 43:   Tie an Overhand Knot in all four cords at once, cut the cords flush with the knot and saturate the end with glue.

FAT BAND ANKLET, BRACELET OR CHOKER

Sample Length: 13 inches
Materials: (2) 7 foot lengths of 45 lb. test hemp
                 (6) 6 foot lengths of 45 lb. test hemp

This piece can also be wrapped twice around the wrist and worn as a bracelet.

Step 1:   Fold the two 7 foot cords in half and tie a Slide Loop Clasp (shown as tied, with loop closed).
Step 2:   Tie 4 Square Knots (a flat sinnet).
Step 3:   Add two of the 6 foot lengths of 45 lb. hemp as new sets of knotters (2 new knotters on each side). To do this, place the midpoint of the cords horizontally across the carriers and use the original knotters to tie a Square Knot to secure the new cords, leaving the ends out to the side.
Step 4:   Divide the 8 cords in the middle, creating two sets of four cords, one set on the right and one set on the left. Use the outside cords in each set as knotters and tie a Square Knot around the two inner cords in each set (one knot on the right and then a knot on the left). This creates two Square Knots, side by side.
Step 5:   On *each side* (right and then left), add two more 6 foot lengths of 45 lb. cord as new knotters. To do this, place the midpoint of the cords horizontally across the carriers and use the knotters from Step 4 to tie a Square Knot to secure the new cords. This adds 4 new knotters to each side and gives a total of 16 cords in the piece.
Step 6:   Use the 4 new inside or center cords (two from the right half and two from the left half) as knotters and carriers and tie 2 Square Knots in the center of the piece. Using the four cords on either side of the knots just tied, tie a Square Knot in each of these sets. This gives three sets of knots side by side.
Step 7:   Now divide the 16 cords, from right to left, into 4 sets of four cords each. If clarification is needed: The four cords on the right consist of the two outside (right) cords added in Step 5 and one knotter and one carrier from the right set in Step 6. The middle right set consists of one knotter and one carrier from the right set in Step 6 and one knotter and one carrier from the center set in Step 6. The middle left set consists of one knotter and one carrier from the center set in Step 6 and one knotter and one carrier from the left set in Step 6. The left set consists of one knotter and one carrier from the left set in Step 6 and the two outside cords (left) added in Step 5.
Step 8:   Tie 3 Square Knots in each of the outside sets created in Step 7. Tie 2 Square Knots in each of the middle sets created in Step 7. There should be four columns of Square Knots, side by side; the bottoms of these columns should be at about the same level.
Step 9:   Begin tying an Alternating Square Knot pattern (see **Page 15** if needed). This involves dividing the cords into sets of two. Set the outside sets to either side for the first row. Going from right to left, use two cords from each of the four columns in Step 8 to tie a single Square Knot. There should be 3 Square Knots, side by side to complete the first row.
Step 10: For the second row, again split the cords into four sets of four. Use all 16 cords to tie four Square Knots in this row. Repeat Step 8 but tie only one Square Knot in each set.

FAT BAND ANKLET, BRACELET OR CHOKER (Continued)

Step 11:   Repeat Steps 9 and 10.

Step 12:   Repeat Step 9 and 10 again, for a total of six rows of alternating single Square Knots.

Step 13:   Repeat Step 9, but tie 2 Square Knots (instead of one) with each set of cords.

Step 14:   Repeat Step 10, but again tie 2 Square Knots (instead of one) with each set of cords.

Step 15:   Repeat Steps 13 and 14.

Step 16:   Repeat Steps 13 and 14 again.

Step 17:   Repeat Step 13, for a total of seven rows of Alternating double Square Knots.

Step 18:   Repeat Step 14, but return to tying just one Square Knot with each set of cords.

Step 19:   Repeat Step 13, but again return to tying just one Square Knot with each set of cords.

Step 20:   Repeat Steps 18 and 19.

Step 21:   Repeat Steps 18 and 19 again, for a total of six more rows of Alternating single Square Knots.

Step 22:   Tie the next row with two Square Knots in each set of cords (as in Step 14). There should be four columns of 2 Square Knots each.

Step 23:   Tie one more Square Knot in each of the outside columns (right and left).

Step 24:   Now it is time to begin gathering the cords together and taper the Fat Band back down to a single column. As the extra cords are brought together over the next several steps, keep in mind that it is best to cut them out at different lengths so that the knotting will taper smoothly down until there are only two knotters and two carriers remaining. Remember to apply glue to all cuts.

Step 25:   Bring the two outer cords from each of the middle columns into the outside sets (6 cords in each set) and tie a Gathering Knot on each side. The Gathering Knots should be two Square Knots tied around four carriers.

Step 26:   Use the four cords remaining in the center (two from each of the middle columns) to tie two Square Knots.

Step 27:   Divide the cords from the middle column (Step 25) and put two into the right set of cords and two into the left set of cords. Tie two Square Knots in each set (right and left).

Step 28:   Bring all the cords together in a single set and tie a Gathering Knot using the two outside cords as knotters. This Gathering Knot should consist of two Square Knots. Remember to gradually cut out the extra cords as the pattern continues and glue each cut end.

Step 29:   Tie two more Square Knots.

Step 30:   This choker has a knuckled end, which provides adjustable sizing. Begin tying this now by tying an Overhand Knot in each of the knotter cords (to the side). This provides a hook for the slide loop clasp to catch on.

Step 31:   Tie four Square Knots, then add another set of Overhand Knots, one on each side of the choker.

Step 32:   Repeat Step 31 until the desired length is reached.

Step 33:   Add a Switch Knot for balance with the beginning of the pattern.

Step 34:   Add at least three more Square Knots.

Step 35:   To end, cut the carriers, add glue to the cuts and tie at least two more Square Knots. Tie an Overhand Knot to the side in each of the knotters, cut the knotters and glue the Overhand Knots.

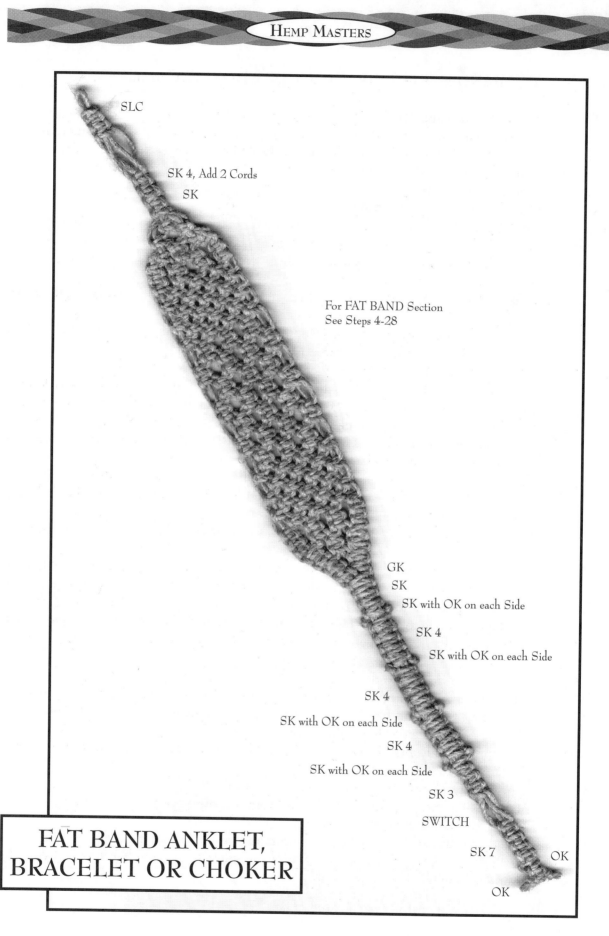

SLC

SK 4, Add 2 Cords

SK

For FAT BAND Section
See Steps 4-28

GK

SK

SK with OK on each Side

SK 4

SK with OK on each Side

SK 4

SK with OK on each Side

SK 4

SK with OK on each Side

SK 3

SWITCH

SK 7

OK

OK

FAT BAND ANKLET, BRACELET OR CHOKER

SLC

HEMP-CATCHING DREAMS:
CAR MIRROR CHARM

LSK 4

SWITCH

LSK 3

SWITCH

LSK 5

BEAD   LSK 2, Cut 1 Carrier

RHK 8

BEAD
RHK 5

BEAD
RHK 5

BEAD, RHK
LSK 2   A

B

H

C

See Steps 17-29
for Hoop

G

D

E   F

HEMP-CATCHING DREAMS: CAR MIRROR CHARM

Sample length: 9 inches plus 3 inch metal hoop
Materials: (2) 5 foot lengths of 45 lb. test hemp
    (2) 2.5 foot lengths of 45 lb. test hemp
    (1) 1.5 foot length of 45 lb. test hemp
    (4) 8 mm trade or bone beads (white in the example)
    (1) 3 inch metal hoop

Hook the Slide Loop Clasp around the bead to the side to create a loop to hang around your car's rear view mirror. May all your Hemp Dreams come true!

Step 1:   Fold the two 5 foot cords in half and tie a Slide Loop Clasp (shown as tied, with loop closed).
Step 2:   Tie a Left Square Knot Sinnet which is 4 Square Knots long (a flat sinnet).
Step 3:   Tie a Switch Knot, which is about an inch long.
Step 4:   Tie 3 Left Square Knots.
Step 5:   Tie another Switch Knot.
Step 6:   Tie 5 Left Square Knots.
Step 7:   Place a bead on *one* of the bead carriers. Bring the same carrier around the bead, then around itself above the bead. Still using the same carrier, tie an Overhand, with the bead in the loop. The bead should be snug against the square knot sinnet already tied. This will force the bead above the plane of the rest of the knotting (turned to the side in the photo for clarity).
Step 8:   Place the carrier from Step 7 back in it's normal position and tie a Left Square Knot in line with the rest of the knots in the sinnet. Do *not* attempt to wrap the knotters around the bead in any way. Tying this knot in the regular manner will secure the other side of the bead snug against the sinnet.
Step 9:   Tie 1 more Left Square Knot, then cut the carrier which did not go through the bead. Put some glue on the cut end. There are now two knotters and one carrier.
Step 10:   Tie 8 Right Half Knots (a spiral sinnet).
Step 11:   String one of the beads on the single carrier. Bring the knotters around either side of the bead and tie a Right Half Knot snug against the bottom of the bead.
Step 12:   Add 4 more Right Half Knots.
Step 13:   String a third bead on the carrier. Bring the knotters around either side of the bead and tie a Right Half Knot snug against the bottom of the bead.
Step 14:   Add 4 more Right Half Knots.
Step 15:   String the fourth bead on the carrier. Bring the knotters around either side of the bead and tie a Right Half Knot against the bottom of the bead.
Step 16:   Add 2 Left Square Knots.
Step 17:   Take the 3 inch metal hoop and one of the 2.5 foot lengths of cord. Leave a 2 inch tail and wrap the cord around the metal ring until all the metal is hidden by wrapped hemp. Finish with the two ends of the wrapping cord next to one another. Tie a Left Square Knot with the two ends to secure the wrapping in place.
Step 18:   Use the wrapping cord ends as carriers and add them to the hanging strap of the charm

HEMP-CATCHING DREAMS: CAR MIRROR CHARM (Continued)

(see **Page 8**). Leave about a half inch of cord(s) between the hoop and the last knot tied in the hanging strap. Use the original knotters to tie these wrapping cords in place as carriers with an extra tight Left Square Knot.

Step 19: Cut all the extra carriers (leaving only the wrapping cords from the hoop) and put some glue on the cut ends, and tie a tight Left Square Knot over the glued ends. Continue tying Left Square Knots down to the wrapped ring if any space remains between the knotters and the ring.

Step 20: Wrap the Knotters around the hoop and cross them over one another. The right knotter should now be on the left and the left knotter should now be on the right. Tie two Square Knots over the last knots tied above the hoop. Cut the knotters and put glue on the cut ends and on the last two Square Knots.

Step 21: Take the second 2.5 foot cord and attach one end to the wrapped ring with glue and a Double Half Hitch, about 1 1/4 inch to the right of the hanging strap (**A**).

Step 22: Attach the cord to the hoop a second time, right next to the hanging strap (**B**). Use a Double Half Hitch and glue. Leave a loop in the cord which is about 1 3/4 inches long (just long enough to reach 1/4 of the distance to the center when pulled in that direction).

Step 23: Attach the cord to the hoop at six more locations, spaced evenly around the hoop (for a total of 8 attachments; **A-H**). Use Double Half Hitches and glue at each location. Leave loops of cord between each attachment point which are the same length (about 1 3/4") as the loop in Step 22. Cut off any excess cord.

Step 24: Attach the 1.5 foot cord to the hoop at the same point as the last attachment point (**H**) in Step 23. Again, use a Double Half Hitch and glue.

Step 25: Use this new cord to pull the outer loops to the center of the hoop. Start by wrapping this cord around the center of the loop between the first and second attachment points (**A & B** - see Step 22). Pull the cord snug against the outer loop. Wrap the new cord around the center of the loop between the second and third attachment points (**B & C**), leave a loop of about a half inch and pull it snug against the second outer loop.

Step 26: Continue around the hoop, pulling the outer loops towards the center with the new cord, and leaving secondary loops of about a half inch.

Step 27: After pulling the loop between the seventh and eighth attachment points (**G & H**) to the center, reverse direction and begin pulling the secondary loops to the center. Leave a half inch loop in the new cord and wrap the cord around the center of the last secondary loop created.

Step 28: Continue back around the loop, wrapping the cord around the center of every other secondary loop. Finish by wrapping the cord around the center of the loop created when reversing direction in Step 27. Pull the cord snug and adjust the looping so the openings within the hoop are as consistent as possible. Note: If one of the "skipped" secondary loops does not pull tight as this third pass is made, it may be necessary to wrap the cord around its center to pull it tight (as was done in the example pictured). Remember, each piece is unique.

Step 29: Take the end of the cord back out to the rim of the hoop and tie it off with a Double Half Hitch and glue. Choose a point between the first and last original attachment points (**A & H**) which pulls the whole pattern tight and looks good. Make sure the cord is taught throughout the interior of the hoop before tying off this cord.

PEACE TO YOU: CAR MIRROR CHARM

Sample length: 12 inches
Materials:
    (2) 8 foot lengths of 45 lb. test hemp
    (9) 1.5 foot lengths of 45 lb. test hemp
    (7) 1 foot lengths of 180 lb. test hemp twine (fillers)
    (5) 8 mm trade beads (4 dark blue and one red in the example)
    (1) long, narrow crystal, 0.5" to 0.75" long (quartz in the example)

Hooking the Slide Loop Clasp around the bead on the end of the added strap is an easy way to hang this charm around your car's rear view mirror. Peace To You Too!

Step 1:   Fold the two 8 foot cords in half and tie a Slide Loop Clasp (shown on right, with loop closed).

Step 2:   Tie a Square Knot Sinnet which is 7 Square Knots long ((A) a flat sinnet).

Step 3:   Tie a short Switch Knot, which is about a half inch long.

Step 4:   Tie 3 Square Knots.

Step 5:   Add 2 of the 1.5 foot cords across the existing carriers (see **Page 8**). Leave them *out to the sides* and secure them in place with 3 Square Knots (tied with the original knotters and carriers). This adds 4 new cords, two on either side of the original cords.

Step 6:   Using *only* the four cords *just added*, tie a Square Knot Sinnet which is the same length as the section already tied (including the slide loop clasp). After the first Square Knot, cut one of the carriers and glue the cut end. This creates a second knotted section (**B**), connected to the first section only at the point where the 4 cords were added. This new section should be approximately 3 3/4 inches long (23 Square Knots in the example).

Step 7:   Place a bead (red) on the bead carrier of the new knotted section. Bring the knotters around either side of the bead, and tie a Square Knot snug against the other side of the bead to secure it in place.

Step 8:   Tie a second Square Knot, then cut and glue the carriers. Take each knotter individually and tie an Overhand Knot on either side of the piece, snug against the sides of the second Square Knot. Glue these knots as well, then cut off any excess cord.

Step 9:   Return to the original knotters and carriers and tie 3 Right Half Knots (a spiral sinnet (**C**)).

Step 10:   Add 2 more of the 1.5 foot cords across the existing carriers and tie a Half Knot, with the original knotters, snug around the new cords.

Step 11:   Pinch the new cords (4; 2 on either side) parallel to the existing carriers and tie a tight Square Knot around all six cords with the original knotters. Stop knotting with the original knotters for the time being (leave them out to either side).

Step 12:   Tie one Square Knot around the hidden new cords (4), using the original carriers as knotters.

Step 13:   Take the 7 180 lb. test filler cords and lay the center of the filler cords across the bead carriers. Again use the original carriers as knotters. Bring them over all 7 filler cords and tie a tight Square Knot around the four new cords. Leave the filler cords and the

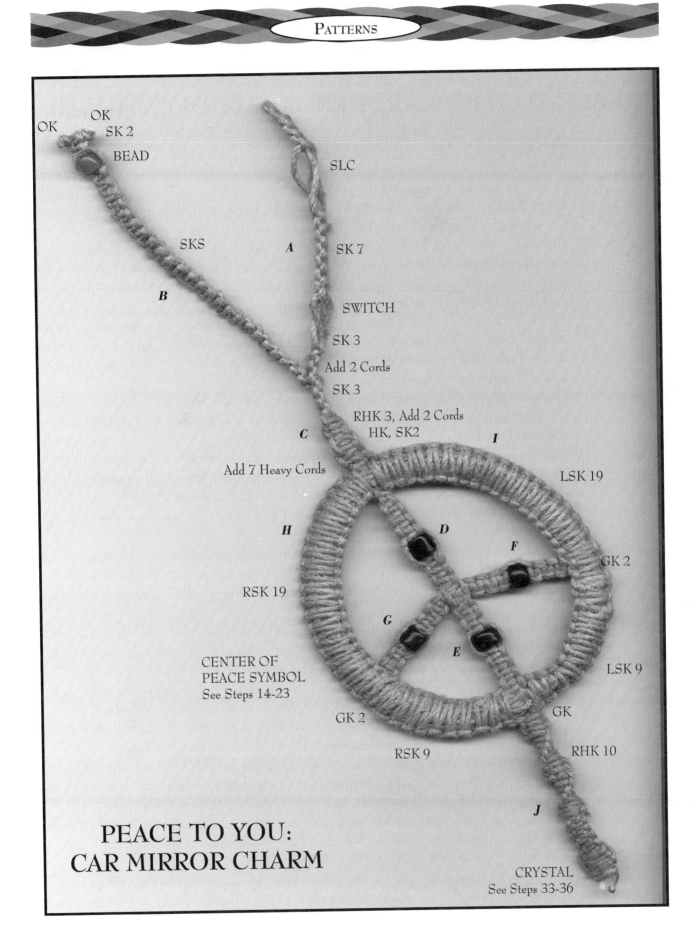

OK

OK

OK SK 2

BEAD

SLC

SKS

A

SK 7

B

SWITCH

SK 3

Add 2 Cords

SK 3

RHK 3, Add 2 Cords
HK, SK2

C

I

Add 7 Heavy Cords

LSK 19

D

F

H

GK 2

RSK 19

G

E

CENTER OF
PEACE SYMBOL
See Steps 14-23

LSK 9

GK 2

GK

RSK 9

RHK 10

J

PEACE TO YOU:
CAR MIRROR CHARM

CRYSTAL
See Steps 33-36

PEACE TO YOU: CAR MIRROR CHARM (Continued)

original carriers out to either side for the time being. These and the original knotters will be used to create the border of the peace sign.

Step 14: Using the four new cords as knotters and carriers, begin tying the vertical, inner arm of the peace sign (**D**). Start with 4 Right Square Knots.

Step 15: Still using the new cords, string one bead (dark blue) on the carriers. Bring the knotters around either side of the bead and tie a Right Square Knot against the bottom of the bead.

Step 16: Add 2 more Right Square Knots.

Step 17: Add 4 more of the 1.5 foot cords across the existing carriers. Leave all four *out to the sides* and secure them in place with a Right Half Knot, followed by 2 Right Square Knots (tied with the current knotters and carriers). This adds 8 new cords, four on either side of the current cords (**E**).

Step 18: Continue the vertical arm by stringing a second bead (dark blue) on the carriers. Bring the knotters around either side of the bead and tie a Right Square Knot against the bottom of the bead.

Step 19: Add 3 more Right Square Knots. Leave these cords as they are for the time being.

Step 20: Next tie the right and left inner arms of the peace symbol. Use the four cords added to either side of the vertical arm in Step 17 as knotters and carriers. Begin by tying 3 Left Square Knots on one side of the vertical arm (**F**). The two side arms are identical, so it doesn't matter which side is tied first.

Step 21: On the same arm, string one bead (dark blue) on the carriers. Bring the knotters around either side of the bead and tie a Left Square Knot against the bottom of the bead.

Step 22: Add 2 more Left Square Knots to this arm and leave these cords as they are for the time being.

Step 23: Repeat Steps 20 through 22 on the other side of the vertical arm to add the second side arm of the peace symbol (**G**). Again, leave these cords as they are for the time being.

Step 24: Now it is time to create the border of the peace symbol. Return to the top of the vertical inner arm. Use the original knotters above the fillers and the original carriers below the fillers as the new knotters. This gives one top knotter and one bottom knotter on either side of the vertical inner arm. Tie the left side of the border from top to bottom (**H**) and then tie the right side from top to bottom (**I**).

Step 25: Begin by tying 19 Right Square Knots around the fillers. This creates a sinnet which is approximately 3 inches long. Gradually curve the fillers towards the bottom of the vertical arm as these knots are tied.

Step 26: Add the four cords from the left inner side arm to the fillers and tie 2 Right Square Knots around the fillers (a Gather Knot) and the cords from the side arm. Cut the four cords from the side arm and apply glue to the cut ends.

Step 27: Tie 9 more Right Square Knots around the fillers (and cut ends until they run out). The bottom portion of the sinnet should be approximately 2 inches long. Leave these cords

PEACE TO YOU: CAR MIRROR CHARM (Continued)

as they are for the time being.

Step 28: Repeat Steps 25 through 27 on the right side of the symbol to complete this side of the border. Tie *Left Square Knots* on this side of the border (replacing the right square knots on the first side). Gather the right side arm cords into the border after 19 Left Square Knots have been tied and then complete the bottom portion of the sinnet border.

Step 29: Cut the filler cords on both sides of the border and butt the ends against one another (interlace them slightly). These ends should come together beneath the center of the inner vertical arm. Bring the four cords from the inner vertical arm around the filler cords from top to bottom (two in front and two in back of the border).

Step 30: Take the two knotters on the inside edge of the border (one right and one left) and cross them on the back side of the border. Tie a Square Knot (a Gather Knot) around the four cords of the vertical inner arm, just below the outside edge of the border. Use this knot to pull the butted ends of the filler together.

Step 31: Place the knotters from Step 30 parallel to the four cords from the vertical inner arm, making all six cords carriers. Use the two knotters from the outside edge of the border (one right and one left) to tie a Right Half Knot around these six cords.

Step 32: Using these same knotters, tie 9 more Right Half Knots around the six cords gathered in Step 31 ((J) a spiral sinnet). Leave these cords as is for the time being.

Step 33: Use the last 1.5 foot cord to tie Square Knots around the crystal. Leave about 1/4" of the crystal tip exposed. Place the crystal at the center of the cord, bring the ends around the crystal about 1/4" from the tip, and begin tying Square Knots, using the crystal as a carrier. Continue knotting until the end of the crystal is reached.

Step 34: Add the knotters from the crystal to the carriers from Step 32. The tip of the crystal should point down and the new carriers from the crystal should point up. Leave about a half inch of carriers between the spiral sinnet and the end of the crystal. Use the knotters from Step 32 to tie a Right Half Knot around all 8 carriers. Fold the ends of the two new carriers (from the crystal) down and tie a Left Half Knot around the folded ends and the rest of the carrier cords.

Step 35: Cut the carriers to various lengths between the last knot and the crystal. Glue the cut ends and tie Square Knots around them until the end of the crystal is reached. Use the same knotters as in Step 34.

Step 36: Bring the knotters around either side of the crystal and tie a Square Knot around the crystal, just above the first knot already on the crystal. Cut the knotters and glue the cut ends.

KARMA'S LOVE: CAR MIRROR CHARM

Sample length: 13.25 inches
Materials: (2) 8 foot lengths of 45 lb. test hemp
                    (2) 3 foot lengths of 45 lb. test hemp
                    (2) 1 foot lengths of 45 lb. test hemp
                    (8) 1 foot lengths of 20 lb. test hemp twine
                    (3) buttons (1 red two-hole and 2 black with back loop in the example)
                    (1) large charm bead (bronze heart in the example)

Hooking the Slide Loop Clasp around the bead on the end of the added strap is an easy way to hang this charm around your car's rear view mirror. May your Yin and Yang always be in harmony!

Step 1:   Fold the two 8 foot cords in half and tie a Slide Loop Clasp (shown on right, with loop closed).
Step 2:   Tie a Square Knot Sinnet which is 7 Square Knots long ((A) a flat sinnet).
Step 3:   Add a 1 foot and a 3 foot cord across the existing carriers. Leave them *out to the sides* and secure them in place with 3 Square Knots (tied with the original knotters and carriers). This adds 4 new cords, two on either side of the original cords (one long and one short on either side).
Step 4:   Using *only* the four cords *just added*, tie 3 Square Knots. Use the two short cords as carriers. After the second Square Knot, cut out one of the carriers and glue the cut end. This creates a second knotted section (B), connected to the first section only at the point where the 4 cords were added.
Step 5:   Continuing with the second knotted section, tie 14 Right Half Knots (a spiral sinnet).
Step 6:   Add Square Knots (about 7) until this section is the same length as the original section (including the slide loop clasp).
Step 7:   Place a button (black) on the bead carrier of the new knotted section. Bring the knotters around the back side of the button and tie a Square Knot snug against the opposite side of the button holes or back loop to secure it in place.
Step 8:   Tie 3 more Square Knots, then cut and glue the carriers. Take each knotter individually and tie an Overhand Knot on either side of the piece, snug against the sides of the last Square Knot. Glue these knots as well, then cut off any excess cord.
Step 9:   Return to the original knotters and carriers and tie 3 Right Half Knots (C).
Step 10:   Add 1 of the 1 foot lengths of 20 lb. test twine as a new carrier. Use 3 Square Knots, tied with the original knotters, to secure the new cord. This smaller cord allows a wider selection of beads and buttons to be used.
Step 11:   Stop knotting with the original knotters for the time being (leave them out to either side).
Step 12:   Use the original carriers to tie a Square Knot around the new 20 lb. test carrier.
Step 13:   Take the 7 remaining 20 lb. test cords and lay the center of these filler cords across the single thin bead carrier. Again use the original carriers as knotters. Bring them over all 7 filler cords and tie a tight Square Knot around the bead carrier. Leave the filler cords

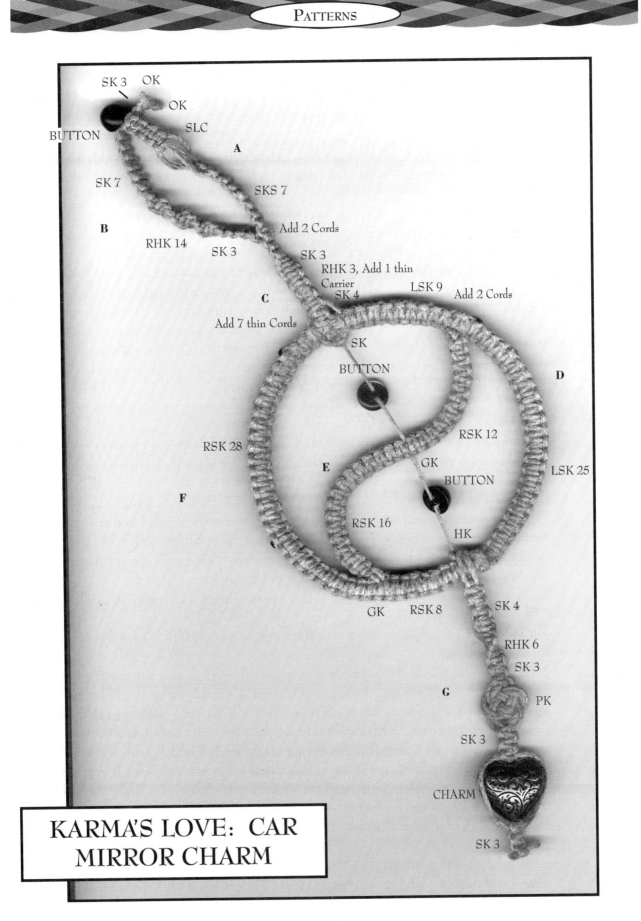

SK 3 OK

OK

BUTTON SLC

A

SK 7 SKS 7

B Add 2 Cords

RHK 14 SK 3 SK 3

RHK 3, Add 1 thin
Carrier LSK 9 Add 2 Cords

SK 4

C

Add 7 thin Cords

SK

BUTTON D

RSK 28 RSK 12

GK

E BUTTON LSK 25

F RSK 16

HK

GK RSK 8 SK 4

RHK 6

SK 3

G PK

SK 3

SK 3

CHARM

SK 3

KARMA'S LOVE: CAR MIRROR CHARM

KARMA'S LOVE: CAR MIRROR CHARM (Continued)

and the original carriers out to either side. The original carriers and the knotters will be used to create the border of the Yin Yang symbol. Leave the new carrier as it is for the time being.

Step 14: Use the original knotters above the fillers and the original carriers below the fillers as the new knotters. This gives one top knotter and one bottom knotter on either side of the fillers. Tie the right side of the border from top to bottom (D), then tie the center arm (E) and finally tie the left side of the border from top to bottom (F).

Step 15: Begin by tying 9 Left Square Knots around the fillers, to the right side of the original knotted section. This creates a sinnet which is approximately 1 1/2 inches long. Gradually curve the fillers towards the bottom of the symbol as these knots are tied.

Step 16: Place the centers of a 3 foot and a 1 foot cord across the fillers (at right angles). Leave them out to the sides and tie a Half Knot around the fillers and the new cords. This gives four new cords. Pull the new cords which are outside the symbol around the fillers to the inside of the circle being formed. These cords will be used to tie the inner arm, but leave them as they are for now.

Step 17: Tie 25 more Left Square Knots around the fillers (D). The bottom portion of the sinnet should be approximately 5 inches long. Leave these cords as they are for the time being.

Step 18: Return to the new cords added in Step 16. Begin tying the inner arm (E), using the short cords as carriers and the longer ones as knotters. First, tie 12 Right Square Knots.

Step 19: Take the button or bead (red) chosen for the top yin of the symbol and string it on the single 20 lb. bead carrier at the top of the symbol (from Steps 10-13). Lay this 20 lb. carrier across the two short carriers in the inner arm. This carrier should cross the inner arm and the end should reach to the bottom of the symbol.

Step 20: Bring the inner arm knotters around this crossing cord and secure it in place with a Right Square Knot.

Step 21: Tie 15 more Right Square Knots with the inner arm cords (E). Leave this section as is for the time being.

Step 22: Next tie the left side of the border from top to bottom (F). Use the remaining original knotter above the fillers and the remaining original carrier below the fillers as the new knotters. Tie *Right Square Knots* on this side of the border. Begin by tying 28 Right Square Knots (about a 5 inch sinnet) around the fillers on the left side of the symbol.

Step 23: Gather the inside arm cords into the border after the 28 Right Square Knots have been tied (place them parallel to the fillers and secure them in place with the border knotters). Then complete the bottom portion of the sinnet border (another 8 Right Square Knots, including the knots which secure the inner arm cords).

Step 24: Cut the filler cords on both sides of the border and butt the ends against one another (interlace them slightly). These ends should come together beneath the center of the original knotted section.

Step 25: Take the button or bead (black) chosen for the bottom yang of the symbol and string it

KARMA'S LOVE: CAR MIRROR CHARM (Continued)

on the single 20 lb. bead carrier hanging down from the middle of the inner arm. Bring this carrier down between the ends of the filler cords.

Step 26: Using the two knotters on the inside edge of the border (one right and one left), tie a Half Knot around the piece of 20 lb. twine just above the fillers.

Step 27: Pull the knotters from Step 26 down the back side of the border and cross them. Now use them as carriers. Use the two knotters from the outside edge of the border (one right and one left) to tie a Square Knot around the three carrier cords (one 20 lb. and two 45 lb.).

Step 28: Pull on the 20 lb. cord to adjust the inner arm and vertical twine into the Yin and Yang position shown in the photograph. Tie 3 more Square Knots with the knotters from Step 27. Cut the thin carrier cord and glue the cut end.

Step 29: Using these same knotters, tie 5 or 6 Right Half Knots around the two remaining carrier cords.

Step 30: Tie 3 Square Knots (G).

Step 31: Add one Pretzel Knot.

Step 32: Tie 2 Square Knots, cut one of the carrier cords and glue the cut end. Tie one more Square Knot.

Step 33: String the large charm bead (heart) on the single carrier. Bring the knotters around the sides of the bead and tie a Square Knot snug against the bottom of the bead.

Step 34: Tie one more Square Knot and cut and glue the carrier.

Step 35: Tie a final Square Knot, cut the knotters and glue the cut ends.

Variation: Beads can be added around the border by periodically stringing a bead on one of the fillers and continuing as normal. These beads will sit above the plane of the border knots.

SIMPLE LACE AND OTHER BEADED JEWELRY PATTERNS:
For Ages 7 To 70
by Mary Ellen Harte

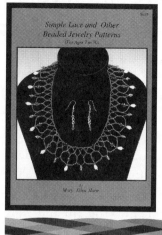

This is the first seed bead workbook, written with beaders of all ages in mind, to supply large, clear, graphic illustrations as well as instructions on how to thread many popular jewelry patterns. This presentation provides the framework for creating attractive and colorful necklaces, rings, bracelets, earrings and napkin holders. The patterns come from diverse cultures throughout the world, many of which were collected by the author during her travels. This is a wonderful new book for beaders of all ages and skills!

CRAFT CORD CORRAL:
Bead Stringing Projects For Everyone
by Janice S. Ackerman

New and innovative crow bead stringing projects that utilize craft cord, leather cord, imitation sinew or hemp. When these projects are embellished with conchos, feathers and other finery, they represent an entirely new look in jewelry and home decorative items. The complete step-by-step directions are easy to follow and well illustrated with clear, concise drawings. Projects are easily mastered by beginners, yet the professional results attract experienced crafters as well. Great book with fun projects!

TRADITIONAL INDIAN CRAFTS
by Monte Smith

This book includes complete, illustrated instructions on all the basics of Leather, Bone and Feather crafts of the American Indian. Projects include: Bear Claw Necklace, Sioux Dancing Choker, Quilled Medicine Wheels, Feathered Dance Whistle, Double-Trailer Warbonnet and much more. In addition, craft techniques described include quill wrapping, shaping feathers, how to size projects, using imitation sinew, how to antique bone, obtaining a "finished" look and hints on personalizing craft projects. Designed so even beginners can create authentic Indian crafts with ease!

PLAINS INDIAN And MOUNTAIN MAN ARTS & CRAFTS
by Charles W. Overstreet

This illustrated handbook is an exciting exploration of the arts, crafts and accoutrements made and used by the Plains Indians and Mountain Men in the early 1800's. Employing traditional and modern methods, this complete how-to manual features 45 projects ranging from rawhide to an Arapaho saddle. Using the easy-to-follow instructions and illustrations, re-creation of these historical items becomes a simple task. A varied selection of items that can be made relatively inexpensively is provided. The author provides interesting historical background on the use and significance of each piece.

CLASSIC EARRING DESIGNS
by Nola May

No wonder beading is so popular! It's easy to learn, and creating beautiful personalized accessories is a very satisfying experience. This collection of Comanche Weave (or Brick Stitch) earring designs has color combinations inspired by Mother Nature herself. Aimed at beginning and intermediate beaders, this book has 52 new and exciting patterns that are sure to stimulate the creativity of advanced beaders as well. There are easy-to-follow instructions, lots of illustrations and sections on materials, techniques, basic earring patterns, and variations on the basic designs. A great way to learn how to bead!

TECHNIQUES OF NORTH AMERICAN INDIAN BEADWORK
by Monte Smith

This informative, easy to read book contains complete instructions on every facet of beadwork. Included is information on selecting beads; materials and their use; designs; making looms and loomwork; applique stitches such as the lazy, "crow", running, spot and return stitches; bead wrapping and peyote stitch; making rosettes and beaded necklaces; and beadwork edging. There is a selected bibliography and an index. The book features examples and photos of beadwork from 1835 to the present time from 23 tribes. Anyone interested in Native American craftwork will profit from owning this book.